SpringerBriefs in Molecular Science

Green Chemistry for Sustainability

Series editor

Sanjay K. Sharma, Jaipur, India

More information about this series at http://www.springer.com/series/10045

Süheyda Atalay · Gülin Ersöz

Novel Catalysts
in Advanced Oxidation
of Organic Pollutants

Springer

Süheyda Atalay
Faculty of Engineering
Ege University
İzmir
Turkey

Gülin Ersöz
Faculty of Engineering
Ege University
İzmir
Turkey

ISSN 2191-5407   ISSN 2191-5415   (electronic)
SpringerBriefs in Molecular Science
ISSN 2212-9898
SpringerBriefs in Green Chemistry for Sustainability
ISBN 978-3-319-28948-9   ISBN 978-3-319-28950-2   (eBook)
DOI 10.1007/978-3-319-28950-2

Library of Congress Control Number: 2015960783

Printed on acid-free paper

This Springer imprint is published by SpringerNature
The registered company is Springer International Publishing AG Switzerland

Preface

Many industrial activities utilize organic compounds as a key chemical in the production of chemicals; hence, a wide range of organic compounds is detected in industrial effluents. Advanced oxidation processes (AOPs), characterized by the generation of highly reactive free radicals, which can unselectively oxidize and mineralize organic compounds, are widely used for the treatment of industrial wastewaters. Using catalyst is a key strategy in these treatment processes since they change the operating conditions and also enhance the treatment efficiency. Hence, there is an ever-present need for developing new environmentally friendly methods toward "greener" catalyst production processes.

The main goals are technologies of the invention, design and application of catalysts and processes to reduce or to eliminate the use and generation of hazardous substances, and where possible utilize renewable raw materials. This book will offer comprehensive overview of the most recent developments in catalysts used for advanced oxidation of organic pollutants. The novel catalysts used in AOPs will be mainly investigated in two categories: homogeneous and heterogeneous catalysts. Among these catalyst categories, it will be mainly focused on nanocatalysts, perovskite-type catalysts, and green catalysts used in several types of advanced oxidation processes, such as those based on Fenton chemistry and photocatalytic oxidation, or hybrid technologies that combine different processes. Topics outlined will be described in terms of both catalyst preparation and characterization and reaction chemistry, and process technology.

This book also will present case studies highlighting the application of advanced oxidation processes for wastewater treatment using the above-mentioned catalysts.

Contents

About the Authors

Süheyda Atalay is a Professor in the Chemical Engineering Department, Faculty of Engineering, Ege University. She completed her B.Sc. and M.Sc. degrees at the Chemical Engineering Department, Faculty of Engineering, Ege University, in 1974 and 1976, respectively. She has been awarded Ph.D. in 1984. She has been working at the Chemical Engineering Department of Ege University since 1975 with different positions starting from research assistant to full-time professor. She has published over 40 papers in international journals, and she has oral and poster communications at international conferences. She has been teaching in different areas of chemical engineering, particularly in subjects related to reaction engineering, chemical engineering design, modeling and simulation, chemical engineering economics, kinetics and catalysis, principles of heterogeneous catalysis, hazardous waste treatment, multiphase reactors, principles of heterogeneous catalysis, and optimization in chemical reactors. She has a considerable experience of Treatment of Gas and Liquid Pollutants, Advanced Oxidation Processes, Catalyst Preparation and Characterization and Catalytic Reactions. Besides the teaching responsibilities, she has also been in charge of many administrative positions such as Head of the Department, Member of Executive Administrative Committee, and recently Dean of the Faculty. At university level, Prof. Atalay is coordinating institutionally both international (ERASMUS) and national (FARABI, MEVLANA) mobility programs. She has been responsible for the implementation of ECTS/DS studies at Ege University. She is a member of the Turkish National Team of Bologna Experts since 2004. She is the National Academic Contact Person of Turkey for ECTS/DS issues. She has a great experience on the accreditation of engineering programs, and she is one of the program evaluators of engineering accreditation body in MÜDEK (Association for

Evaluation and Accreditation of Engineering Programmes), Turkey. She has been taken part in the thematic network projects Teaching and Research in Engineering Education (TREE) and European and Global Engineering Education Academic Network (EUGENE) which were supported by the European Commission.

 **Gülin Ersöz** is an assistant Professor at the Faculty of Engineering in Chemical Engineering at Ege University. She received her bachelor and master degrees in Chemical Engineering at Ege University in 1999 and 2002, respectively. She has been awarded Ph. D. in 2009. She has been working at the Chemical Engineering Department of Ege University since 1999. Her research interests are mainly wastewater treatment by advanced oxidation processes and catalyst preparation and characterization. She has several publications on generally wastewater treatment in international peer-reviewed journals and oral and poster communications at international conferences. She has been teaching reaction engineering, hazardous waste treatment, chemical engineering economics, and differential equation courses. Ersöz is the co-coordinator of international (ERASMUS) and the coordinator of national (MEVLANA) mobility programs of the Chemical Engineering Department.

Chapter 1
Introduction

Abstract In recent days, disposal of wastewater streams containing highly toxic organic pollutants generated by variety of industrial processes is a major environmental problem. The effluent discharged into the environment must have no or at least minimum impact on human and environmental health. The increase in the disposal of refractory organics demands for newer technologies for the complete mineralization of these wastewaters for human health and environmental needs. The introductory section presents a brief background on the wastewater treatment. Throughout the chapter some conventional treatment methods are introduced and mainly their drawbacks are mentioned. Subsequently, principles of current Advanced Oxidation Processes (AOPs) are considered since they are highly competitive water treatment technologies for the removal of those organic pollutants not treatable by conventional techniques due to their high chemical stability and/or low biodegradability.

1.1 Wastewater Treatment

Wastewater treatment is a process to convert wastewater into an effluent that can be either returned to the water cycle with minimal environmental issues or reused.

The choice of the appropriate method must be carried out considering several factors, both technical such as treatment efficiency or plant simplicity and economical (Herney-Ramirez et al. 2010). Many methods have been applied conventionally for the treatment of wastewater discharged from various industries. These conventional treatment techniques comprise physical methods (e.g. screening filtration, sedimentation, aeration, flotation and skimming, degasification, equalization), chemical methods (e.g. chlorination, neutralization, coagulation, adsorption, ion exchange) and biological process (e.g. activated sludge, aerobic and anaerobic digestion, trickling filtration, oxidation ponds, lagoons, septic tanks) (Gaehr et al. 2008). All these methods can be used separately or combined with other processes to increase the overall treatment efficiency (Herney-Ramirez et al. 2010).

© The Author(s) 2016   1
S. Atalay and G. Ersöz, *Novel Catalysts in Advanced Oxidation
of Organic Pollutants*, SpringerBriefs in Green Chemistry for Sustainability,
DOI 10.1007/978-3-319-28950-2_1

Some of the most common conventional treatment methods are briefly introduced below:

Filtration

Filtration is the process of separating suspended solid matter from a liquid, by causing the latter to pass through the pores of some substance, called a filter. The liquid which has passed through the filter is called the filtrate.[1]

Filtration methods such as ultrafiltration, nanofiltration and reverse osmosis have been used for wastewater treatment, chemical recovery and water reuse. In the textile industry, filtration methods can be applied for recycling of pigment rich wastewaters in addition to mercerizing and bleaching wastewaters. The specific temperature and chemical composition of the wastewaters determine the type and porosity of the filter to be used (Verma et al. 2012).

Sedimentation

Sedimentation is the tendency for particles in suspension to settle out of the fluid in which they are entrained, and come to rest against a barrier. This is due to their motion through the fluid in response to the forces acting on them: these forces can be due to gravity, centrifugal acceleration or electromagnetism. Sedimentation process is particularly applied when the wastewater contains of high percentage of settable solids or when the waste is subjected to combined treatment including sewage. The sedimentation tanks are designed to settle the small and light particles under gravity. The settled sludge is discharged from the sedimentation tanks by mechanical scrapping into hoppers and pumping it out subsequently.

Membrane technology

It is well known that the utilization of membrane technology for wastewater treatment is effective. Nevertheless, membrane filtration processes suffers from several significant problems such as fouling and subsequent clogging of membrane surfaces, which resulted in reduction of wastewater flows, shorter membrane life leading to increase the cost of the treatment. For these reasons, membrane filtration processes may not be suitable to treat the wastewaters especially having a high-strength of organic matters (Yoon et al. 2014).

Coagulation–flocculation

In the coagulation-flocculation processes suspended solids agglomerate together into bigger bodies so that an easier physical filtration processes can be applied to remove them. These processes are usually followed by gravity separation (sedimentation or flotation) and filtration. A chemical coagulant, such as polymers or iron and aluminum salts, is added to wastewater to facilitate bonding among particulates. Coagulants

---

[1]http://www.lenntech.com/chemistry/filtration.htm, 2015

create a chemical reaction and eliminate the negative charges which cause particles to repel each other.[2] Coagulation of the dyes has limitations because some highly soluble low molecular weight cationic dyes might not be removed effectively and the main disadvantage of this process is need for the disposal of the sludge produced inherently (Rodrigues et al. 2013).

Ion exchange

In ion exchange processes, ions interchange reversibly between the solid and liquid phases, where a resin (an insoluble substance) removes ions from an electrolytic solution and releases other ions of like charge in a chemically equivalent quantity without any change in the structure of the resin (Kurniawan et al. 2006). The main negative drawback of the ion exchange method is the cost of the refreshing the resins.

Activated sludge process

In the activated sludge process, air and a biological floc composed of bacteria are used to treat the wastewater. Even though it is an economical process, due to the high amount of hydrophobic compounds present in the industrial effluent, the biological stage of the plant often suffers from sludge bulking and foaming problems. Also, a tertiary treatment stage is required to remove refractory compounds (Lotito et al. 2014).

Adsorption

Mainly, adsorption is a mass transfer phenomenon by which a substance is transferred from the liquid or gas phase to the surface of a solid, and bounds by physical or chemical interactions. Different types waste materials such as, saw dust, alum sludge, mud, nut shells and hulls, and other waste materials can be used as adsorbents. Because of its large surface area and high adsorption capacity, adsorption using activated carbon can remove pollutants effectively (Chiban et al. 2012). However, adsorption is ineffective against some types of pollutants, the regeneration of the adsorbent is expensive and results in loss of the adsorbent, non-destructive process (Crini 2006).

As mentioned above, the typical conventional treatment methods have some of the disadvantages such as:

- high cost,
- an apparent low cost option is offered by the biological oxidation, but the organic pollutant/s has/have to be biodegradable, diluted and of low toxicity,
- transfer of the aqueous organics to another phase, leaving the contaminants undestroyed (Herney-Ramirez et al. 2010).

---

[2]http://www.drinking-water.org/html/en/Treatment/Coagulation-Flocculation-technologies.html, 2015

To overcome these disadvantages of conventional treatment methods and since the organic pollutants in water are a great concern for the environment, especially toxic and non-biodegradable pollutants that cannot be treated by conventional methods, various chemical oxidation methods have arisen. Among these processes, the advanced oxidation processes appear to be the most promising methods.

1.2   Advanced Oxidation Processes

Advanced oxidation processes (AOPs), which involve highly reactive oxygen species like hydroxyl radical ($\cdot$OH), have attracted increasing attention due to their potential capability in the removal of recalcitrant organic pollutants.

Advanced oxidation processes (AOPs) are generally non-waste generating technologies which are better alternative wastewater treatments for organic pollutant when common wastewater treatment technologies mentioned above such as sedimentation, adsorption, flocculation, filtration, reverse osmosis are insufficiently effective (Soon and Hameed 2011; Martínez-Huitle and Brillas 2009; Konstantinou and Albanis 2004).

The highly reactive oxygen species, $\cdot$OH radicals, can be generated either by one or the combination of chemical oxidation by using ozone, hydrogen peroxide with or without the radiation assisted sources such as ultrasounds, ultraviolet, and etc. (Soon and Hameed 2011).

New, economically viable, more effective methods for pollution control and prevention are required for environmental protection.

Advanced oxidation processes (AOPs) are known to have an outstanding priority over other wastewater treatment methods, due to their ability to mineralize the organic pollutants.

Advanced oxidation processes are technologies based on the generation of highly reactive species, the hydroxyl radicals, used in oxidative degradation procedures for organic pollutants in aquatic media. They mainly depend on specific chemical reactions which are characterized by the generation of chemical oxidizing agents capable of degrading the pollutants.

The efficiency of the AOP can be increased by the use of an appropriate catalyst. Use of economically available and environmentally friendly catalysts for pollutant oxidation, which can also be regenerated, will certainly increase the treatment efficiency. Generally by using the appropriate catalyst, the problem of harmful intermediates can be easily countered ending at complete oxidation of the pollutants.

Consequently, the objective of this book is to provide the readers with the knowledge of novel catalysts in advanced oxidation of organic pollutants in aqueous solutions. In the view of the importance of using catalysts in advanced oxidation, it is the purpose of this chapter:

- to present an overview on wastewater treatment
- to investigate and introduce the most important aspects of various AOPs
- to discuss the key factors in green chemistry and catalysis
- to give some information on catalysis in advanced oxidation processes
- to present a perspective to the readers about the researches performed in literature on catalytic advanced oxidation processes

These issues will be discussed these in mainly three sections:

- Green Chemistry and Catalysis
- Advanced Oxidation Processes
- Catalysis in Advanced Oxidation Processes

Thus the main topic will be covered from its basic premises to its implementation in practice. Finally the concluding remarks will be given.

References

Chiban M, Soudani A, Sinan F, Persin M (2012) Wastewater treatment by batch adsorption method onto micro-particles of dried Withania frutescens plant as a new adsorbent. J Environ Manage 95:S61–S65

Crini G (2006) Non-conventional low-cost adsorbents for dye removal: a review. Bioresour Technol 97(9):1061–1085

Gaehr F, Hermanuts F, Oppermann W (2008) Ozonation— an important technique to comply with new German laws for textile wastewater treatment. Water Sci Technol 30(3):255–263

Herney-Ramirez J, Vicente MA, Madeira LM (2010) Heterogeneous photo-Fenton oxidation with pillared clay-based catalysts for wastewater treatment: a review. Appl Catal B 98:10–26

Konstantinou IK, Albanis TA (2004) TiO$_2$-assisted photocatalytic degradation of azo dyes in aqueous solution: kinetic and mechanistic investigations: a review. Appl Catal B 49:1–14

Kurniawan TA, Chan GYS, Lo W-H, Babel Sandhya (2006) Physico–chemical treatment techniques for wastewater laden with heavy metals. Chem Eng J 118:83–98

Lotito AM, De Sanctis M, Di Iaconi C, Bergna G (2014) Textile wastewater treatment: aerobic granular sludge vs activated sludge systems. Water Res 54:337–346

Martínez-Huitle CA, Brillas E (2009) Decontamination of wastewaters containing synthetic organic dyes by electrochemical methods: a general review. Appl Catal B 87:105–145

Rodrigues CSD, Madeira LM, Boaventura RAR (2013) Treatment of textile dye wastewaters using ferrous sulphate in a chemical coagulation/flocculation process. Environ Technol 34(6):719–729

Soon AN, Hameed BH (2011) Heterogeneous catalytic treatment of synthetic dyes in aqueous media using Fenton, and photo-assisted Fenton process. Desalination 269:1–16

Verma AK, Dash R, Bhunia P (2012) A review on chemical coalugation/flocculation technologies for removal of colour from textile wastewaters. J Environ Manage 93:154–168

Yoon Y, Hwang Y, Kwon M, Jung Y, Hwang T-M, Kang J-W (2014) Application of O$_3$ and O$_3$/H$_2$O$_2$ as post-treatment processes for color removal in swine wastewater from a membrane filtration system. J Ind Eng Chem 20(5):2801–2805

Chapter 2
Green Chemistry and Catalysis

Abstract Green chemistry has always been a lively research field. In the time period, the emphasis of catalysis research has significantly shifted and spread from traditional applications in green chemistry. This chapter gives an introduction to overall theme of green chemistry and catalysis emphasizing the concepts such as homogeneous and heterogeneous catalysts, preparation and characterization of catalysts. Readers will find coverage of some important types of green catalysts namely nanocatalysts and perovskite type catalysts with an emphasis on their preparation and characterization. The section on catalyst preparation is concerned with the preparation of bulk catalysts and supported catalysts, with an emphasis on general principles. For the supported catalysts the relation between the method of preparation and the surface chemistry of the support is highlighted. The section on catalyst characterization summarizes the most common techniques in four subtitles. Structural Analysis, Thermal Analysis, Spectroscopic Techniques and Microscopic Techniques.

2.1 Green Chemistry

Green chemistry also named as sustainable chemistry is defined as the practice of chemical science and manufacturing in a manner that is sustainable, safe, and non-polluting and that consumes minimum amounts of materials and energy while producing little or no waste material (Sheldon et al. 2007).

Anastas and Warner also defined Green Chemistry as *"The invention, design and application of chemical products and processes to reduce or to eliminate the use and generation of hazardous substances"* (Anastas and Warner 1998).

The design of environmentally benign chemicals and processes are guided by the 12 Principles of Green Chemistry developed by Anastas and Warner.

These 12 principles of green chemistry can be described in detail as follows:

© The Author(s) 2016   7
S. Atalay and G. Ersöz, *Novel Catalysts in Advanced Oxidation
of Organic Pollutants*, SpringerBriefs in Green Chemistry for Sustainability,
DOI 10.1007/978-3-319-28950-2_2

1. Prevention
   It is better to prevent waste than to treat or clean up waste after it has been
   created.
2. Atom Economy
   Synthetic methods should be designed to maximize the incorporation of all
   materials used in the process into the final product.
3. Less Hazardous Chemical Synthesis
   Wherever practicable, synthetic methods should be designed to use and generate substances that possess little or no toxicity to human health and the
   environment.
4. Designing Safer Chemicals
   Chemical products should be designed to affect their desired function while
   minimizing their toxicity.
5. Safer Solvents and Auxiliaries
   The use of auxiliary substances (solvents, separation agents, etc.) should be
   made unnecessary wherever possible and innocuous when used.
6. Design for Energy Efficiency
   Energy requirements of chemical processes should be recognized for their
   environmental and economic impacts and should be minimized. If possible,
   synthetic methods should be conducted at ambient temperature and pressure.
7. Use of Renewable Feedstocks
   A raw material or feedstock should be renewable rather than depleting when-
   ever technically and economically practicable.
8. Reduce Derivatives
   Unnecessary derivatization (use of blocking groups, protection/deprotection,
   temporary modification of physical/chemical processes) should be minimized
   or avoided if possible, because such steps require additional reagents and can
   generate waste.
9. Catalysis
   Catalytic reagents (as selective as possible) are superior to stoichiometric
   reagents.
10. Design for Degradation
    Chemical products should be designed so that at the end of their function they
    break down into innocuous degradation products and do not persist in the
    environment.
11. Real-Time analysis for Pollution Prevention
    Analytical methodologies need to be further developed to allow for real-time,
    in-process monitoring and control prior to the formation of hazardous
    substances.
12. Inherently Safer Chemistry for Accident Prevention
    Substances and the form of a substance used in a chemical process should be
    chosen to minimize the potential for chemical accidents, including releases,
    explosions, and fires.

These principles have been used as guideline and criteria for the scientists and hence the researches are mainly focusing around the principles of green chemistry. Consequently, it can be said that the modern society has an ever increasing demand for environmentally friendly processes.

It is well known that catalysts have an enormous impact on the chemical industry because they enable reactions to take place and make reaction processes both more efficient and environmentally friendly. A suitable catalyst can enhance the rate of a thermodynamically feasible reaction but cannot change the position of the thermodynamic equilibrium. Most catalysts are solids or liquids, but they may also be gases (Julkapli and Bagheri 2015).

2.2 General Considerations on Catalysts: Homogeneous versus Heterogeneous Catalysts

Catalysis is generally divided into two types, homogeneous and heterogeneous. Heterogeneous catalysis is where the catalyst and the reactants are in the different physical phases, while homogeneous is where both are in the same phase.

2.2.1 Homogeneous Catalysis

The field of homogeneous catalysis can be characterized as a source of easily arranged, selective catalysts with high activity.

In homogeneous catalysis, the catalyst is in the same phase as the reactants and products. In general, homogeneous catalysts exhibit excellent catalytic activities with high selectivities in particular. But, it is difficult for the homogeneous catalyst to be separated from reaction media.

2.2.2 Heterogeneous Catalysis

Heterogeneous catalysts also offer many advantages, some of which are not displayed by their homogeneous counterparts, including recyclability, ease of separation from the reaction mixture and use in continuous flow processes. It is highly desirable to develop new systems that blend the many advantages of heterogeneous catalysis with the versatility of homogeneous catalysts.

Heterogeneous catalysts have several advantages compared to other catalytic processes (Tang 2007):

- they avoid formation of inorganic salts
- they are regenerable

- easy to handle, safe to store and has long life time
- easy and inexpensive of recovery and recycling
- the selectivity and activity of homogeneous catalysts under mild reaction conditions is unbeaten by their heterogeneous counter parts.

2.2.3 Green Catalysts

Recently, growing attention is being directed towards the development of innovative catalytic systems with high performance from the point of environmentally greener processes, economical efficiency and minimum consumption of resources. The application of catalysis to reduce toxicity and renewable energy systems, and efficiency makes it a central focus area for green chemistry research.

Green Catalysis is a subtitle of green chemistry but the most important one and one of the urgently needed challenges facing engineers now is the design and use of environmentally benign catalysts. Green and sustainable catalyst should possess, higher activity, higher selectivity, efficient recovery from reaction medium, durability or recyclability, cost effectiveness.

In recent years the development of catalysts for processes to replace conventional ones has made a significant contribution to the reduction of environmental pollutants. Thus, there is an increasing interest on the topic of green catalysis recently. It not only includes developing new catalysts which can offer stable, highly effective catalytic performances, but considers the application of environmentally friendly catalyst preparations.

Numerous studies have been focused on green catalysts. In this book, among these catalyst categories, it will be mainly focused on nanocatalysts, perovskite type catalysts and green catalysts used in several types of advanced oxidation processes.

2.2.3.1 Nanocatalysts

Nanomaterials have structured components with at least one dimension less than 100 nm. Nanomaterial is expected to be fruitful area for green chemistry catalysis due to the increasing ability to design in nano state and the high surface areas found in nano materials.

Employing green chemistry principles for the production of nanoparticles can lead to a great reduction in waste generation, less hazardous chemical syntheses, and an inherently safer chemistry in general (Bhattacharya et al. 2013).

Materials reduced to the nanoscale can show different properties compared to what they exhibit on a macroscale, enabling unique applications. For instance, stable materials turn combustible (aluminum); solids turn into liquids at room temperature (gold); insulators become conductors (silicon). A material such as gold, which is chemically inert at normal scales, can serve as a potent chemical catalyst at nanoscales (Chaturvedi et al. 2012).

Nanostructured materials are potential candidates for the innovative catalyst because of the unique properties such as enormous surface areas they exhibit compared to their bulk counterparts (Glaser 2012).

The nano sized particles increase the exposed surface area of the active component of the catalyst, thereby enhancing the contact between reactants and catalyst dramatically and mimicking the homogeneous catalysts. However, their insolubility in reaction solvents renders them easily separable from the reaction mixture like heterogeneous catalysts, which in turn makes the product isolation stage effortless. Also, the activity and selectivity of nano-catalyst can be manipulated by tailoring chemical and physical properties like size, shape, composition and morphology (Polshettiwar and Varma 2010).

In the emerging regime of nano-catalysis the synergetic effect of the nanosized catalyst is known to be important for the overall performance, where global processes such as the transport of the reactant atoms to the catalyst could play an important role in the overall reaction kinetics (Xie et al. 2013). In general, it can be concluded that due to rapid ·OH radical based oxidation reactions, AOP by nanocatalysts is characterized by high reaction rates and hence short treatment times. In addition it is known that less toxic intermediate products are formed and pollutants can be degraded in ppb level.

Nanotechnology can step in a big way in lowering the cost and hence become more effective than recent techniques for the removal of pollutants from water in the long run. In this perspective nanoparticles can be used as potent sorbents as separation media, as catalysts for photochemical destruction of contaminants; nanosized zerovalent iron used for the removal of metals and organic compounds from water (Bhattacharya et al. 2013).

Advances in nanoscale science and engineering suggest that many of the current problems involving water quality could be resolved or diminished by using nanosorbents, nanocatalysts, bioactive nanoparticles, nanostructured catalytic membranes, nanotubes (Bhattacharya et al. 2013).

Nanomaterial properties desirable for water and wastewater applications include high surface area for adsorption, high activity for (photo)catalysis, antimicrobial properties for disinfection and biofouling control, superparamagnetism for particle separation, and other unique optical and electronic properties that find use in novel treatment processes and sensors for water quality monitoring (Qu et al. 2013).

Nanosized metal oxides, including nanosized ferric oxides, manganese oxides, aluminum oxides, titanium oxides, magnesium oxides and cerium oxides, are classified as the promising ones for wastewater treatment systems (Hua et al. 2012).

Consequently, the catalysts that are prepared by methods of nanotechnology used in advanced oxidation processes are of particular interest because of their environmentally friendly features.

2.2.3.2 Perovskites

The most of the catalysts used in modern chemical industry are based on mixed metal oxides. Among the mixed metal oxides, perovskite-type oxides are attracting much scientific application interest owing to their low price, adaptability, and thermal stability, which often depend on bulk and surface characteristics (Guiotto et al. 2015; Gupta et al. 2015).

In its ideal form, perovskites are described by the generalized expression ABX_3. They have a cubic structure, where each of the cubes consists of three different chemical elements A, B and X present in a 1:1:3. Atoms A and B are metal cations (positively charged ions) and the X atoms are not metal anions (negatively charged ions, usually oxygen). The cation A is the largest atomic radius is the hub center, the cation B occupies the eight vertices and in the center of the edges of the cubic cell are centered anions X (Yazdanbakhsh et al. 2011).

Although the most numerous and most interesting compounds with the perovskite structure are oxides, some carbides, nitrides, halides, and hydrides also crystallize in this structure.

Perovskite materials exhibit many unusual properties that may also furnish practical applications. Such phenomena as high magnetoresistance, ferroelectricity, superconductivity, charge ordering, spin-dependent transport, high thermopower and the interleaving of structural, magnetic and transport properties are those typically observed from this family of materials. Perovskites with transition metal ions (TMI) on the B site show an enormous variety of intriguing electronic or magnetic properties. This variety is not only related to their chemical flexibility, but also and to a larger extent related to the complex character that transition metal ions play in certain coordination with oxygen or halides (Lemmens and Millet 2004).

The Perovskite-type catalysts are promising candidates for the advanced oxidation of wastewater in environmental catalysis applications because:

- wide range of elements that can be incorporated into the structure, combination of elements with different oxidation states.
- high temperature resistance.
- it is a good alternative as it is possible to produce metal particles high dispersion, thus reducing the formation of coke in reaction.

2.3 Fundamentals in Catalyst Preparation

The catalytic process is realized on the catalyst surface. Therefore, the catalyst surface should be as large as possible, but moreover, surface must be accessible to the reactants.

The design of a catalyst covers all aspects from choice of the active phases to the method of forming the particles.

The desired structures of the catalysts and the criteria for a good catalyst are

- homogeneity
- often nano-structured, but not necessarily
- adapted pore structure
- uniform particle size distribution
- suitable shape and mechanical stability
- activity
- selectivity
- cost

and generally all these structures depend on the preparation method.

Generally, the catalysts may be classified according to the preparation procedure as: bulk catalysts or supports and impregnated catalysts. According to the preparation methods the catalytic active phase is generated as a new solid phase or the active phase is introduced or fixed on a pre-existing solid by a process which intrinsically depends on the support surface (Campanati et al. 2003).

2.3.1 Basic Preparation Techniques for Bulk Catalysts and Supports

2.3.1.1 Precipitation and Co-precipitation

In the process the desired component is precipitated from the solution. The precipitation process is used for preparation of bulk catalysts and support material. Co-precipitation is used for simultaneous precipitation of more than one component. Catalysts based on more than one component can be prepared easily by co-precipitation.

In the co-precipitation procedure the solutions containing the metal salt and a salt of a compound that will be converted into the support are contacted under stirring with a base in order to precipitate as hydroxides and/or carbonate. After washing, these can be transformed to oxides by heating (Pinna 1998).

Metal salt solutions and support precursors are combined together, and then calcined. It was widely used method, due to simplicity, economic and reproducibility.

Co-precipitation method offers some advantages. These are:

1. produces nanoparticles in large quantities in a relatively short amount of time
2. utilizes inexpensive and readily available chemicals as precursors
3. easy control of particle size and composition and
4. various possibilities to modify the particle surface state and overall homogeneity.

2.3.1.2 Sol–Gel

Sol–gel processing is one of the routes for the preparation of porous materials by solidification (without precipitation) from a solution phase.

The reaction proceeds in two steps:

-M-O-R + H_2O == -M-OH + ROH (hydrolysis)
-M-OH + XO-M == M-O-M + XOH (condensation)

The sol–gel methods show promising potential for the synthesis of mixed oxide catalysts. The versatility of the sol–gel techniques allows control of the texture, composition, homogeneity, low calcination temperatures (minimizing the undesired aggregation of the particles), and structural properties of solids, and makes possible production of tailored materials such as dispersed metals, oxidic catalysts and chemically modified supports (Brinker and Scherer 1990; Prasad and Singh 2011).

The main advantages of sol–gel technique are versality, low temperature process and flexible rheology allowing easy shaping. The procedure involving sol–gel technique is as follows (Kandasamy and Prema 2015).

- preparation of homogeneous solution by dissociation of metal organic precursor in the organic solvent or in organic salt solution,
- transformation of precursor oxide into a highly cross linked solid,
- hydrolysis leads to sol, dispersion of colloidal particle in liquid done by suitable reagents (generally water),
- further condensation leads to gel, a rigid inter connected organic network.

Citric acid (CA) complexing approach

Citric acid (CA)-assisted sol–gel method (namely Pechini approach) is a facile synthesis for producing homogeneous nanocomposites. In this method the citric acid is used as chelating agent to ensure the formation of homogeneous transparent metal-citrate gels (Wei and Hua 2007; Prasad and Singh 2011).

Alkoxide sol–gel method (Pechini Method)

The Pechini method based on polymeric precursors, is used to prepare spinels and it does not require high temperature calcinations. The method has good stoichiometric control as well as reproducibility. This method consists of the formation of a polymeric resin between a metallic acid chelate and polyhydroxide alcohol by polyesterification. The method involves relatively easy synthesis route when compared to the other conventional processes. Low operating temperature and control over the end stoichiometry are the main advantages of this technique (Lessing 1989; Pimentel et al. 2005; Prasad and Singh 2011).

Non-alkoxide sol–gel route

Non-alkoxide sol–gel process, involving hydrolysis and condensation of metal salts, avoids the disadvantage of alkoxide sol–gel process (high sensitivity to moist

environment), however, has still the disadvantage of different hydrolysis suscepti-
bilities of the individual components (Cui et al. 2005). One of the advantages of this
method is the important reduction of the required calcination temperatures, minimizing the undesired aggregation of the particles (Prasad and Singh 2011).

2.3.1.3 Solvothermal Reaction

In this reaction, precursors are dissolved in hot solvents. If the solvent is water then
the process is referred to as hydrothermal method. Solvent other than water can
provide milder and friendlier reaction conditions.

2.3.1.4 Solid-State Reaction

The solid-state reaction route is the most widely used method for the preparation of
polycrystalline solids from a mixture of solid starting materials. Solids do not react
together at room temperature over normal time intervals and it is necessary to heat
them to much higher temperatures (1000–1500 °C) in order for the reaction to occur
at an appropriate rate.

The factors on which the feasibility and rate of a solid state reaction depend
include, reaction conditions, structural properties of the reactants, surface area of
the solids, their reactivity and the thermodynamic free energy change associated
with the reaction (West 2005).

2.3.1.5 Flame Spray Pyrolysis

Flame spray pyrolysis is a novel one step method for preparation of especially
nano-sized particles. In the process a liquid feed metal precursor dissolved in an
organic solvent is sprayed as micrometer sized droplets with an oxidizing gas into a
flame zone. The spray is combusted and the precursor is converted into nanosized
metal or metal oxide particles, depending on the metal and the operating conditions
(Høj 2012).

2.3.2 Basic Preparation Techniques for Supported Catalysts

Many industrial catalysts consist of metals or just as a carrier but it may actually
contribute catalytic activity. Further, the interaction between the active phase and
the support phase can affect the catalytic activity

2.3.2.1   Impregnation

Impregnation method appears to be simple, economic and able to give a repro-ducible metal loading which is however limited by the solubility of the metal precursor. In the procedure the support is contacted with a certain amount of solution of the metal precursor, usually a salt (e.g. metal nitrate, chloride), then it is aged, usually for a short time, dried and calcined.

According to the amount of solution used, there are two types of impregnation: incipient wetness or wet impregnation.

In the wet type of impregnation an excess amount of solution with respect to the pore volume of the support is to be used. The system is left to age for a certain time under mixing, filtered and dried (Pinna 1998).

In incipient wetness impregnation, the volume of the solution of appropriate concentration is equal or slightly less than the pore volume of the support. The maximum loading is limited by the solubility of the precursor in the solution (Campanati et al. 2003).

Ion exchange

Ion exchange process consists of replacing an ion in an electrostatic interaction with the surface of a support by another ion species. The support containing ions (X) is immersed into an excess volume (compared to the pore volume) of a solution con-taining other ions (Y). Ions Y gradually penetrate into the pore space of the support, while ions X pass into the solution, until equilibrium is reached (Campanati et al. 2003).

2.3.2.2   Deposition–Precipitation

This method includes two stpes:

*   precipitation from bulk solutions or from pore fluids and
*   interaction with the support surface.

The dissolution of the metal precursor is followed by adjustment of the pH to achieve a complete precipitation of the metal hydroxide which is deposited on the surface of the support. The hydroxide formed is subsequently calcined and reduced to the elemental metal (Campanati et al. 2003). The main problem is to allow the precipitation of the metal hydroxide particles inside the pores of the support: therefore the nucleation and growth on the support surface will result in a uniform distribution of small particles on the support (Pinna 1998).

2.3.2.3   Inert Gas Condensation

The inert gas condensation technique, in which nano particles are formed by the evaporation of a metallic source in an inert gas, had been extensively used to produce fine nano particles.

Methods:

- Physical Vapor Deposition (PVD)-(no catalytic interaction)
- Chemical Vapor Deposition (CVD)-(with catalytic interaction)

Chemical Vapour Deposition (CVD) is a well known process in which a solid is deposited on a heated surface via a chemical reaction from the vapour or gas phase. In thermal CVD, the reaction is activated by a high temperature above 900 °C. A typical apparatus comprises of a gas supply system, a deposition chamber and an exhaust system (Malik and Singh 2010; White et al. 2009).

2.3.2.4 Inert Structured Media

Preparation of particles in structured medium by imposing constraints in form of matrices such as

- zeolites
- layered solids
- molecular sieves
- micelles/microemulsions
- gels
- polymers
- glasses

Imposing the structure in form of matrices the growth kinetics can be slowed down and directed and the size can be limited leading to nanoparticles with well defined structures (Overney 2010).

2.4 Catalyst Characterization

To investigate the fundamental relations between the state of a catalyst and its catalytic properties several approaches can be adopted. By using the appropriate combination of analysis techniques, the desired characterization on the atomic as well as bulk scale is certainly possible.

A heterogeneous catalyst is a composite material, characterised by:

- the relative amounts of different components (active species, physical and/or chemical promoters, and supports)
- shape
- size
- pore volume and distribution
- surface area

In this section some of the characterization techniques that are most commonly used will be discussed.

These techniques will be summarized in four subtitles:

- Structural analysis
- Thermal analysis
- Spectroscopic techniques
- Microscopic techniques

(a) **Structural analysis**

In heterogeneous catalysis, the reaction ocurs at the surface. Catalysis and catalytic surfaces, hence, need to be characterized by reference to their physical properties and by their actual performance as a catalyst. The most important physical properties are those relating to the surface because the catalyst performance is determined by surface parameters.

Surface area—Brunauer, Emmet and Teller (BET) method

In heterogeneous catalysis, since most of the reactions occur at the surfaces, investigating the surface area is very significant. Surface area measurements are mainly performed by Brunauer, Emmet and Teller (BET) method.

The BET theory is based on various hypothesis:

- the adsorption is supposed to be localized on well defined sites; all the sites have the same energy (homogeneous surface) and each of them can only accommodate one adsorbate molecule
- a multilayer adsorption is supposed to occur even at very low pressure. The adsorbed molecules in the first layer acting as adsorption sites for the molecules of the second layer and so on; there is no lateral interaction between adsorbed molecules.
- an adsorption–desorption equilibrium is supposed to be effective between molecules reaching and leaving the solid surface.

The BET equation describes the relationship between volume adsorbed at a given partial pressure and the volume adsorbed at monolayer coverage. BET equation can be written in the form:

$$\frac{V}{V_m} = \frac{C(P/P_0)}{(1 - P/P_0)(1 - P/P_0 + C(P/P_0))}$$

$$\frac{1}{V[(P_0/P) - 1]} = \frac{1}{V_mC} + \frac{C-1}{V_mC}\left(\frac{P}{P_0}\right)$$

where P and P_0 are the equilibrium and the saturation pressure of adsorbates at the temperature of adsorption, V is the adsorbed gas quantity (for example, in volume units), and V_m is the monolayer adsorbed gas quantity. C is the BET constant,

$$C = \exp\left(\frac{E_1 - E_L}{RT}\right)$$

where E_1 is the heat of adsorption for the first layer, and E_L is that for the second and higher layers and is equal to the heat of liquefaction. R is ideal gas constant and T is the absolute temperature.

If a plot of $\frac{1}{V[(P_0/P)-1]}$ versus $\left(\frac{P}{P_0}\right)$ is drawn, it will yield a straight line with slope $\frac{C-1}{V_m C}$ and intercept $\frac{1}{V_m C}$ if the BET model holds true. Knowing slope and intercept, V_m can be calculated which is used for the calculation of specific surface area using the following equation

$$S_g = a\frac{V_m N_A}{V}$$

S: total surface area
N_A: the Avagadro constant
V: molar volume of the adsorbent gas
a: adsorption cross section of the adsorbing species,

Pore analysis—The Barrett-Joyner-Halenda (BJH) method

Since a catalytic reaction occurs at the fluid-solid surface a large interfacial area can be helpful or even essential in attaining a significant reaction rate. This area is provided by a porous structure, the solid contains many fine pores, and the surface of these pores supplies the area needed for the high rate of reaction.

A catalyst that has a large area resulting from pores is called a porous catalyst. Sometimes pores are so small that they will admit small molecules but prevent large ones from entering.

In general, catalyst consists of pore in the range of, micropore, mesopore or macropore depending on the preparation conditions and compositions.

Not all catalyst need the extended surface provided by a porous structure, however, some are sufficiently active so that the effort required to create a porous catalyst would be wasted. For such situations one type of catalyst is the monolithic catalyst. They can be porous or non porous.

The pore analyses consist of determining the average pore size, average pore volume and pore size or pore volume distribution.

The Barrett-Joyner-Halenda (BJH) method for calculating pore size distributions is based on a model of the adsorbent as a collection of cylindrical pores. The theory accounts for capillary condensation in the pores using the classical Kelvin equation, which in turn assumes a hemispherical liquid-vapor meniscus and a well-defined surface tension.

X-Ray Diffraction (XRD)

X-ray diffraction is the most powerful and successful technique for determining the structure of crystals. It also gives some idea regarding crystallinity, crystal grain size, lattice parameters, phase composition, lattice defects etc.

The instrument used for this analysis is called an X-ray diffractometer. In the diffractometer, an X-ray beam of a single wavelength is used to examine the specimens. By continuously changing the incident angle of the X-ray beam, a spectrum of diffraction intensity versus the angle between incident and diffraction beam is recorded.

The major applications of XRD are phase identification and average crystallite size determination.

Phase Identification

The catalysts are generally composed of mixture of several phases. Phase identification is based on the comparison of the diffraction pattern of the specimen with that of pure reference phases or with a database.

Average crystallite size determination

The average crystallite size can be determined by Scherrer formula using elementary line broadening analysis.

$$D = \frac{K\lambda}{\beta Cos\theta}$$

where
D = Crystallite size, Å
K = Crystallite-shape factor
λ = X-ray wavelength, Å
θ = Observed peak angle, degree
β = X-ray diffraction broadening, radian

(b) **Thermal analysis**

Temperature Programmed Reduction (TPR)

Temperature programmed reduction is used to determine the reducibility of the catalysts.

Temperature Programmed Desorption (TPD)

Temperature programmed desorption technique measures the desorbed molecules from the sample surface as function of temperature.

Thermo Gravimetric Analysis (TGA)

In thermogravimetric analysis, the change in mass of samples is monitored with an increase in temperature at specified gas environment and heating rate. TGA characterization is used to determine the thermal stability, content of moisture and volatile material, if any, or decomposition of inorganic and organic material in the catalysts.

Differential Thermal Analysis (DTA)

DTA consists of heating a sample and reference material at the same rate and monitoring the temperature difference between the sample and reference.

(c) **Spectroscopic techniques**

Infra Red Spectroscopy

Infra red spectroscopy is a vibrational spectroscopy as it is based on the phenomenon of absorption of infrared radiation by molecular vibrations. IR spectroscopy gives information about the molecular structure of the materials and bonds. Both inorganic and organic materials can be analyzed. By the help of this, compounds are identified and sample composition can be investigated.

Raman Spectroscopy

Raman spectroscopy is used to analyze the internal structure of molecules and crystals. It is based on scattering phenomenon of electromagnetic radiation by molecules.

(d) **Microscopic techniques**

Scanning Electron Microscopy (SEM)

Scanning electron microscopy is a widely accepted technique to extract structural and chemical information point-by-point from a region of interest in the sample. It is generally employed to examine the surface morphologies of the material at higher magnifications. In SEM, image is formed by focused electron beam that scans over the surface area of specimen.

Transmission Electron Microscopy (TEM)

Transmission electron microscopy is a powerful tool that provides information about the morphology, crystallography and elemental composition for advanced materials. It is also an electron microscopic technique with high resolution to extract structural information.

References

Anastas PT, Warner JC (1998) Green chemistry: theory and practice. Oxford Science Publications, Oxford

Bhattacharya S, Saha I, Mukhopadhyay A, Chattopadhyay D, Ghosh UC, Chatterjee D (2013) Role of nanotechnology in water treatment and purification: potential applications and implications. Int J Chem Sci Technol 3(3):59–64

Brinker CJ, Scherer GW (1990) Sol–Gel science: the phys chem sol–gel proces. Academic Press, Boston

Campanati M, Fornasari G, Vaccari A (2003) Fundamentals in the preparation of heterogeneous catalysts. Catal Today 77(4):299–314

Chaturvedi S, Dave PN, Shah NK (2012) Applications of nano-catalyst in new era. J Saudi Chem Soc 16:307–325

Cui H, Zayat M, Levy D (2005) Sol–Gel synthesis of nanoscaled spinels using sropylene oxide as a gelation agent. J Sol–Gel Sci Technol 35:175–181

Glaser JA (2012) Green chemistry with nanocatalysts. Clean Technol Environ Policy 14:513–520

Guiotto M, Pacella M, Perin G, Iovino G, Michelon N, Natile MM, Glisenti A, Canu P (2015) Washcoating versus direct synthesis of $LaCoO_3$ on monoliths for environmental applications. Appl Catal A 499:146–157

Gupta VK, Eren T, Atar N, Yola ML, Parlak C, Karimi-Maleh H (2015) $CoFe_2O_4@TiO_2$ decorated reduced graphene oxide nanocomposite for photocatalytic degradation of chlorpyrifos. J Mol Liq 208:122–129

Høj M (2012) Nanoparticle synthesis using flame spray pyrolysis for catalysis one step synthesis of heterogeneous catalysts. Doctoral Thesis, Technical University of Denmark

Hua M, Zhang S, Pan B, Zhang W, Lv L, Zhang Q (2012) Heavy metal removal from water/wastewater by nanosized metal oxides: a review. J Hazard Mater 211–212:317–331

Julkapli NM, Bagheri S (2015) Graphene supported heterogeneous catalysts: an overview. Int J Hydrogen Energy 40(2):948–979

Kandasamy S, Prema RS (2015) Methods of synthesis of nano particles and its applications. J Chem Pharm Res 7(3):278–285

Lemmens P, Millet P (2004) Spin—orbit—topology, a triptych, in "quantum magnetism". Springer, Heidelberg

Lessing PA (1989) Mixed-cation powders via polymeric precursors. Am Soc Ceramic Bulletin 68 (5):1002–1007

Malik H, Singh AK (2010) Engineering physics. Tata MCGraw Hill Education Private Limited, New Delhi

Overney R (2010) Nanothermodynamics and nanoparticle synthesis. Lecture Notes

Pimentel PM, Martinelli AE, de Araújo Melo DM, Pedrosa AMG, Cunha JD, da Silva Júnior CN (2005) Pechini synthesis and microstructure of nickel doped copper chromites. Mater Res 8 (2):221–224

Pinna F (1998) Supported metal catalysts preparation. Catal Today 41(1–3):129–137

Polshettiwar V, Varma RS (2010) Green chemistry by nano-catalysis. Green Chem 12:743–754

Prasad R, Singh P (2011) Applications and preparation methods of copper chromite catalysts: a review. Bull Chem React Eng Catal 6(2):63–113

Qu X, Brame J, Li Q, Alvarez PJJ (2013) Nanotechnology for a safe and sustainable water supply: enabling integrated water treatment and reuse. Acc Chem Res 46(3):834–843

Sheldon RA, Arends I, Hanefeld U (2007) Green chemistry and catalysis. WILEY-VCH Verlag GmbH & Co. KGaA, Weinheim

Tang Z-R (2007) Green catalysts preparation using supercritical CO_2 as an antisolvent. Doctoral Thesis, Cardiff University

Wei L, Hua C (2007) Synthesis and characterization of Cu-Cr-O nanocomposites. Solid State Sci 9 (8):750–755

West AR (2005) Solid state chemistry and its applications. Wiley, New York

White RJ, Luque R, Budarin VL, Clark JH, Macquarrie DJ (2009) Supported metal nanoparticles on porous materials. Methods and applications. Chem Soc Rev 38:481–494

Xie W, West DJ, Sun Y, Zhang S (2013) Role of nano in catalysis: palladium catalyzed hydrogen desorption from nanosized magnesium hydride. Nano Energy 2:742–748

Yazdanbakhsh M, Tavakkoli H, Hosseini SM (2011) Characterization and evaluation catalytic efficiency of $La_{0.5}Ca_{0.5}NiO_3$ nanopowders in removal of reactive blue 5 from aqueous solution. Desalination 281:388–395

Chapter 3
Advanced Oxidation Processes

Abstract Various organic pollutants are discharged into the aquatic environment. Most of them are not only toxic but also nonbiodegradable. Hence they cannot be treated by biological wastewater treatment plants or other conventional methods. And consequently a need arises to develop effective methods for the degradation of organic pollutants, either to less harmful compounds or to their complete mineralization. Advanced Oxidation Processes (AOPs) have attracted increasing attention due to their potential capability in the removal of recalcitrant organic pollutants. This chapter presents the general process principles on Advanced Oxidation Processes namely and discusses catalysis in Advanced Oxidation Processes. The usage and mechanism of both homogeneous and heterogeneous catalysts in Fenton Oxidation, Catalytic Wet Air Oxidation and Ozonation are discussed comprehensively.

3.1 General Process Principles

Treatment of organic pollutants in aqueous solutions by advanced oxidation processes has received a great deal of attention in recent years. AOPs generate the highly reactive hydroxyl radical ($\cdot$OH) to degrade the chemicals present in wastewater by attacking the pollutants rapidly and nonselectively (Muruganandham et al. 2014).

AOPs offer various alternative methods for the production of hydroxyl radicals and hence they allow a better compliance with specific wastewater treatment requirements. AOPs generally have end products which are nonhazardous and eco-friendly (Muruganandham et al. 2014).

In order to generate highly reactive intermediates, $\cdot$OH radicals, the AOPs may proceed along one of the two routes given below (Munter 2001; Atalay and Ersöz 2015):

- the use of high energetic oxidants such as ozone and H_2O_2 and/or photons. The generation of hydroxyl radical might possibly be by the use of UV, UV/O_3,

© The Author(s) 2016
S. Atalay and G. Ersöz, *Novel Catalysts in Advanced Oxidation of Organic Pollutants*, SpringerBriefs in Green Chemistry for Sustainability, DOI 10.1007/978-3-319-28950-2_3

UV/H_2O_2, Fe^{+2}/H_2O_2, TiO_2/H_2O_2 and a number of other processes (Mandal et al. 2004).

- the use of O_2 in temperature ranges between ambient conditions and those found in incinerators, such as in Wet Air Oxidation (WAO) processes in the region of 1–20 MPa and 200–300 °C;

In general, AOPs involve two stages of oxidation.

Stage 1: the formation of strong oxidants (mainly hydroxyl radicals)
Stage 2: the reaction of these oxidants with organic contaminants in water.

After the ·OH radical is generated, the radical can attack all organic and inorganic compounds. Due to the nature of the substrate there are three different types of possible attacks (Munter 2001; Siitonen 2007; Karat 2013):
The ·OH radical can

- take a hydrogen atom from the pollutant (alkenes, alcohols etc.).
- add itself to the pollutant (aromatics, olefins, etc.)
- transfer its unpaired electron to other substrates (carbonates, bicarbonates etc.)

3.1.1 Classification of AOPs

The processes are further subdivided into processes whether UV source is used in the process and the formation of strong oxidants can be investigated in two either as homogeneous or heterogeneous.

Table 3.1 shows the classification of AOPs (Mota et al. 2008; Atalay and Ersöz 2015).

Table 3.1 Classification of advanced oxidation processes

Non-photochemical	Photochemical
Homogeneous processes	
Ozonation	Photolysis
Ozonation with hydrogen peroxide (O_3/H_2O_2)	UV/H_2O_2
Fenton (Fe^{2+} or Fe^{3+}/H_2O_2)	UV/O_3
Wet air oxidation (WAO)	UV/O_3/H_2O_2
Electrochemical oxidation	Photo-Fenton (Fe^{3+}/H_2O_2/UV)
Heterogeneous processes	
Catalytic wet air oxidation (CWAO)	Heterogeneous photocatalysis
Fenton catalytic ozonation	

3.1.2 Advantages and Disadvantages of AOPs

The advanced oxidation processes offer several advantages over biological or physical processes (Loures et al. 2013; Sharma et al. 2011):

Advantages of AOPs

- can provide the complete mineralization of pollutants;
- can be used for the treatment of organic pollutants which resistant to other treatments, such as biological processes;
- allow the conversion of recalcitrant compounds and refractory contaminants submitted to biodegradation systems;
- can be used in combination with other processes
- possess strong oxidizing power with rapid reaction rates;
- minimum by-product formation
- does not concentrate waste for further treatment with methods such as membranes.
- does not produce materials that require further treatment
- in many cases, consume less energy

Disadvantages of AOPs

Unfortunately, besides having advantages, they have several disadvantages, too.

- not all processes can be scaled-up to industrial needs
- higher capital and operating costs compared with conventional methods
- in some cases control of oxidant concentration and pH correction are required
- process limitations are also related to pH changes, such as particle aggregation and modification of surface properties of catalysts used in heterogeneous.

3.2 Catalysis in Advanced Oxidation Processes

Homogeneous processes occur in a single phase. These homogeneous advanced oxidation processes use ozone (O_3), hydrogen peroxide and UV (UV/H_2O_2) or Fenton reagent (mixture of H_2O_2 with Fe(II), Fe(III) salt) generate hydroxyl radicals. Some combinations of technologies like photo-Fenton and UV/ozone can also be considered in this group (Loures et al. 2013).

Heterogeneous advanced oxidation processes use catalysts to carry out the degradation of organic pollutants. The term heterogeneous refers to the fact that the contaminants are present in the liquid phase, while the catalyst is in the solid phase.

3.2.1   Fenton Oxidation

One of the most intensively studied AOPs is the process based on the Fenton Oxidation, which has many advantages including high degradation efficiency, benign process, inexpensive materials and general applicability.

Fenton and related reactions involve reactions of peroxides (usually H_2O_2) with iron ion to form active oxygen species. The term "Fenton's Reagent" refers to the mixture of hydrogen peroxide and iron ions. The active oxygen species are able to oxidize organic or inorganic compounds. The active sites derived from iron ions which serve as catalyst to break down the hydrogen peroxide molecules into hydroxyl radicals.

3.2.1.1   Homogeneous Fenton (Fe(II) or Fe(III)/H_2O_2)

When all the Fenton reagents are present in the dissolved phase, the process is categorized as homogeneous Fenton.

In this process, the reaction between dissolved Fe(II) or Fe(III) and H_2O_2 in acidic aqueous medium leads to oxidation of Fe(II) to Fe(III) and hence the hydroxyl radicals ($\cdot$OH) are produced.

Fe(II) or Fe(III) catalyzed decomposition of H_2O_2 in acidic medium mechanism is proposed by Barb et al. (1949, 1951a, b) as the sequence of seven reactions (Yalfani 2011). These reactions are given below:

$$Fe(II) + H_2O_2 \rightarrow Fe(III) + \cdot OH + OH^- \qquad (3.1)$$

$$Fe(III) + H_2O_2 \rightarrow Fe(II) + HO_2\cdot + H^+ \qquad (3.2)$$

$$HO\cdot + H_2O_2 \rightarrow HO_2\cdot + H_2O \qquad (3.3)$$

$$HO\cdot + Fe(II) \rightarrow Fe(III) + OH^- \qquad (3.4)$$

$$Fe(III) + HO_2\cdot \rightarrow Fe(II) + O_2H^+ \qquad (3.5)$$

$$Fe(II) + HO_2\cdot + H^+ \rightarrow Fe(III) + H_2O_2 \qquad (3.6)$$

$$HO_2\cdot + HO_2\cdot \rightarrow H_2O_2 + O_2 \qquad (3.7)$$

This process may still lead to the precipitation of some insoluble species like metal hydroxide which are not a part of the main process. The main reagents or factors which can participate in a homogeneous Fenton reaction are H_2O_2, Fe(II), Fe(III) (Yalfani 2011).

The key features of the homogenous Fenton system are their reagent conditions, i.e. $[Fe^{2+}]$, $[Fe^{3+}]$, $[H_2O_2]$ and the reaction characteristics (pH, reaction time, temperature and the concentration of pollutants).

In homogeneous Fenton oxidation systems, mass transfer limitations are negligible and the readily available iron ion in the reaction medium reacts effectively in the degradation process (Pouran et al. 2014).

In the homogeneous phase, the chemical changes that take place mainly depend on the nature of the interactions of the reacting substances between Fenton's reagents with compounds to be degraded (Neyens and Baeyens 2003; Soon and Hameed 2011).

If the homogeneous Fenton reaction is initiated by $Fe(III)/H_2O_2$, it is known that the reaction rate proceeds slower than for $Fe(II)/H_2O_2$, because, $Fe(III)$ is reduced to $Fe(II)$ before hydroxyl radicals are produced (Yalfani 2011).

However, homogeneous processes are also associated with some major disadvantages which result in limited use of these processes (Pouran et al. 2014):

- pH-dependence of the system (very acidic),
- formation of ferric hydroxide sludge and its removal,
- the generated sludge may prevent UV radiation penetration in photo-Fenton process,
- difficulty in catalyst recovery,
- the cost associated with acidification and subsequent neutralization (this may limit the usage of homogeneous Fenton Oxidation system at industrial scale).

Therefore, the application of heterogeneous Fenton reactions as a possible solution to overcome the drawbacks of homogeneous catalysis has been put into perspectives by many engineers.

3.2.1.2 Heterogeneous Fenton

One of the bottle necks faced in wastewater treatment by homogeneous Fenton processes is that iron must usually be removed from the water at the end of the process. The wastewater cannot be discharged with the iron ions if the concentration is above European Union limits (<2 ppm) (Soon and Hameed 2011; Navalon et al. 2010). In addition to these, there are other drawbacks of the homogeneous Fenton Oxidation such as the formation of the sludge, limited reaction pH range 2.5–3.5 which is also very acidic, H_2O_2 scavenger. These have made the development and usage of heterogeneous catalysts inevitable.

Heterogeneous Fenton Oxidation is one of the advanced oxidation processes which has gained wide spread acceptance for higher removal efficiency of pollutants under wide range of pH compared to homogeneous reactions (Pouran et al. 2014).

Heterogeneous catalysis is found to be a good alternative to homogeneous catalysis in Fenton's oxidation process. The studies in literature show the effectiveness of the heterogeneous catalyst in Fenton's process.

Heterogeneous Fenton processes are mostly preferred because most of the iron remains in the solid phase and hence is easily separated from the treated water.

As mentioned before, homogenous Fenton processes are activated by their active sites either with $Fe(II)$ or $Fe(III)$. But in the case of heterogeneous Fenton processes

Table 3.2 Comparison of homogeneous and heterogeneous catalyzed Fenton reactions under different phenomena

Phenomena	Homogeneous Fenton	Heterogeneous Fenton
Phase	Same phase as the reagents	Solid-liquid phase
Mechanism	Only degradation process	Dual processes of physical absorption and desorption in addition to chemical reaction
pH	Tight acidic pH range	Broad pH range
Sludge treatment	High amount of precipitated as ferric hydroxide sludge	Minimal ferric hydroxide is formed
Catalyst loss	High	Limited
Catalyst recovery	Time consuming and expensive	Easy

active sites can be the surface of iron ions existing in multiple forms of $[Fe(OH)^{+2}]$, $[Fe(H_2O)]^{2+}$, $[Fe(H_2O)_6]^{3+}$, $[Fe_2(OH)_2]^{4+}$, Fepolycation, Fe_2O_3 and α-FeOOH.

In heterogeneous Fenton oxidation, both the physical and chemical changes occur on the surface of the catalyst where the active sites are. On these active sites, adsorption of reactant molecules occurs. At the end of the reaction, the product molecules are desorbed and leave the active sites (Soon and Hameed 2011).

It is well known that heterogeneous Fenton catalysts should be developed so that the researchers will not face with any iron leaching. Iron ions significantly leach from the catalyst during the reaction, which causes a loss in activity and forms extra metal ion pollution. Consequently, the heterogeneous catalyst with high activity and stability at low cost must be produced considering the source of iron, type of support, method of preparation and catalysts pretreatment conditions (Yalfani 2011).

In Table 3.2, the comparison that was made by Soon and. Hameed for homogeneous and heterogeneous catalytic Fenton Reagent oxidation under different phenomena are given (Soon and Hameed 2011).

3.2.2 Catalytic Wet Air Oxidation

Wet air oxidation (WAO) was initially proposed and developed by Zimmermann. WAO is an attractive method for the treatment of waste streams with chemical oxygen demand from 10,000–100,000 mg/L that are too dilute to incinerate but too concentrated for biological treatment (Fu et al. 2015; Kim and Ihm 2011). During WAO, the waste is oxidized into carbon dioxide, water and low molecular weight organic acids at elevated temperatures (150–325 °C) and pressures (0.5–20 MPa) using pure oxygen or air as the oxidant.

The wet oxidation, using oxygen or air to completely oxidize organic compounds to carbon dioxide and water, is a clean process not involving any harmful chemical reagent, however, being a non-catalytic process, requires high temperature and high pressure to achieve a complete oxidation (Liotta et al. 2009).

Consequently, due to the high costs and severe conditions, catalytic wet air oxidation (CWAO) has been developed to decrease the severity of the reaction conditions while increasing oxidation rates.

During WAO of organic pollutants, low molecular weight short chain carboxylic acids that are resistant to degradation under moderate conditions and diminish overall effectiveness of the process are formed. For this concern, catalytic wet air oxidation (CWAO) has gained some attention. Hence, the catalyst should also be able to remove these intermediates so that the complete removal of the substrate can be achieved.

The addition of catalysts decreases the operating conditions (using milder conditions of temperature and pressure), enhances the reaction rate, and shortens the reaction time (Luck 1999; Levec and Pintar 2007).

The solid catalyst offers a further advantage compared with homogeneous catalysis, in principle the catalyst being easily recovered, regenerated and reused.

The use of a catalyst strongly improves the degradation of organic pollutants by using a lesser amount of oxidizing agent and milder conditions of temperature and pressure. Moreover, the catalyst allows the selective removal of a single pollutants or a group of similar pollutants among a complex mixture. In addition to these, in literature it is mentioned that although it varies with type of wastewater, the operating cost of CWAO is about half that of non-catalytic WAO due to milder operating conditions and shorter residence time (Levec 1997).

3.2.2.1 Homogeneous Catalytic Wet Air Oxidation

Early studies on CWAO are mainly focused on the homogeneous catalyst, which are mainly represented by Cu, Fe, Ni, Co and Mn is better. According to the literature (Wang et al. 1993; Lin and Ho 1996; Bi 1999; Liu and Zhou 1998), usually ammonia is added as a stabilizer when Cu^{2+} is used as a homogeneous catalyst, then alkali is added and ammonia is evaporated out after CWAO treatment, so this can precipitate out of copper and residual copper is made a resin treatment for recovery.

Jing et al. reported that, although homogeneous catalyst has high activity and high selectivity it is difficult to recover and easy to drain, which will require a secondary treatment (Jing et al. 2012).

3.2.2.2 Heterogeneous Catalytic Wet Air Oxidation (CWAO)

Although the homogenous catalysts, e.g. dissolved copper salts, are effective, an additional separation step is required to remove or recover the metal ions from the treated effluent due to their toxicity, and accordingly increases operational costs. And in addition, the severe operating conditions and high costs in homogeneous wet air oxidation limit its application in wastewater treatment. Thus, the development of active heterogeneous catalysts has received a great attention. Various solid catalysts including noble metals, metal oxides, and mixed oxides have been widely studied for the CWAO of aqueous pollutants.

There is not a big difference between the reaction mechanism of CWAO and WAO. The introducing the catalyst just enhances the production of free radicals. Reaction mechanism of WAO is generally regarded as free radical reactions as a chain reaction that is divided into three phases: chain initiation, chain transfer and chain termination (Jing et al. 2012).

Heterogeneous catalyst can be investigated in three categories which are non-noble metal catalysts, noble metal catalysts and carbon materials for CWAO (Jing et al. 2012).

Non-noble metal catalysts such as Cu, Mn, Co, Ni, Bi and other metals have the advantage of being cheap. However it is reported that the activity of non-noble metal catalysts is rather low and the active component of non-noble metal catalyst has the tendency of leaching. Noble metal catalyst is generally prepared by Ru, Rh, Pt, Ir, Au, Ag and other precious metal loaded on the support. Though noble metal is expensive, the catalytic activity is better. Components of noble metals are more stable in the CWAO process, so the stability of noble metal catalyst primarily depends on the stability of the carrier. Carbon nanotubes (CNTs) are one-dimensional carbon materials, and have unique chemical and thermal stabilities. It is reported that carbon nanotubes have a large specific surface area, pore structure, good stability, etc. and are promising catalyst supports (Jing et al. 2012).

3.2.3 Catalytic Ozonation

Molecular ozone is fairly unstable in water with a half-life time ranging from few seconds up to few minutes, according to the pH of and has low solubility in aqueous media. These are major drawbacks that reduce considerably the contact time in water and interaction with the organic substrates to be decomposed. Ozone alone often results in incomplete degradation of the pollutants (Shahidi et al. 2015). Shahidi et al. mentioned in their study that significant improvements can be brought by using solid catalysts in ozonation (Shahidi et al. 2015).

The low solubility of ozone in aqueous media has also oriented the studies towards using solid catalysts such as polymers and zeolites (Rubin 2008). In this regard, fixed beds of porous glass or metals, or the use of solid catalysts that also act as adsorbents transition metal ions and mixed oxides are also used (Shiraga et al. 2006).

Catalytic ozonation has received increasing attention due to its higher effectiveness in the degradation and mineralization of organic pollutants. Organics difficult to dissociate by single ozonation can be oxidized by catalytic ozonation at ambient temperature and pressure.

Catalytic ozonation can be both homogeneous catalytic ozonation, which is the metal ions as the catalysts present in the reaction system; and heterogeneous catalytic ozonation, where the main catalysts are metal oxide and metal or metal oxide on supports (Sable 2014).

The catalysts generally used in ozonation studies are salt of transition metals such as Fe^{2+}, Fe^{3+}, Mo^{6+}, Mn^{2+}, Ni^{2+}, Co^{2+}, Cd^{2+}, Cu^{2+}, Ag^+, Cr^{3+} and Zn^{2+} (Shahidi et al. 2015).

3.2.3.1 Homogeneous Catalytic Ozonation

Homogeneous catalytic ozonation is based on ozone activation by metal ions present in aqueous solution. The catalysts usually used are transition metals such as Fe(II), Mn(II), Ni(II), Co(II), Cd(II), Cu(II), Ag(I), Cr(III) and Zn(II), to degrade the organic pollutants in water (Pines and Reckhow 2002; Wu et al. 2008; Xiao et al. 2008; Trapido et al. 2005; Pachhade et al. 2009; Beltran et al. 2005).

During this procedure, a metal ion determines not only the reaction rate, but also selectivity and ozone consumption (Nawrocki and Kasprzyk-Hordern 2010). The mechanism of homogeneous catalytic ozonation is based on an ozone decomposition reaction followed by the generation of hydroxyl radicals. In a possible mechanism to explain the role of homogeneous catalysts, the metal ions accelerate the decomposition of ozone to produce the O_2^-, and then electron transfer of O_2 to O_3 to gain O_3 (Gracia et al. 1995) During homogeneous catalytic ozonation, an initial complex can be formed between the organic molecule and the metal ion, followed by oxidation of the complex by ozone, which leads to the formation of hydroxyl radicals (Nawrocki and Kasprzyk-Hordern 2010).

Homogeneous catalytic ozonation can improve the efficiency of removal of organic pollutants in water, but the disadvantage of this technology is to introduce ions resulting in the secondary pollution and non-possibility to reuse the catalyst, which lead to increase the costs of water treatment.

3.2.3.2 Heterogeneous Catalytic Ozonation

In recent years, heterogeneous catalytic ozonation has received much attention due to its high oxidation potential. The major advantage of a heterogeneous over homogeneous catalytic system is the ease of catalyst retrieval from the reaction media. However, the stability and durability of the catalyst under operating conditions is important. In this case heterogeneous catalysts with higher stability and lower loss can improve the efficiency of ozone decomposition, and can be recycled and reused without any further treatment. Due to these advantages, the heterogeneous catalytic ozonation is used widely in water treatment. The efficiency of the catalytic ozonation process depends on a great extent on the catalyst and its surface properties as well as the pH of the solution that influences the properties of the surface active sites and ozone decomposition reactions in aqueous solutions (Kasprzyk-Hordern et al. 2003; Zhao et al. 2008, 2009).

In heterogeneous catalytic ozonation, catalyst is in a solid form while the reaction may proceed in bulk water or on the surface of the catalyst. In many cases the formation of hydroxyl radicals is expected to be responsible for the catalytic activity. Several solid catalysts have been described in literatures which are effective in ozonation of organic molecules in aqueous solutions. Catalysts such as metal oxides, metals supported on oxides, minerals and activated carbon have been mainly used in ozonation. Catalytic efficiency of metal oxides is depends on their physical (surface area, pore size, and surface charge) and chemical properties (chemical stability, and active surface sites).

A wide variety of catalysts have been tested in heterogeneous ozonation oxidation processes such as metal oxides and metals, carbon-based catalysts, clays, clay minerals and montmorillonite, mixed hydroxide catalysts.

MnO_2, TiO_2, Al_2O_3, Fe_2O_3, WO_3, CuO, CeO_2, Ni_2O_3, CoO, V_2O_5, Cr_2O_3, MoO_3, CeO, and CuO–CeO_2 mixtures represent typical examples of metal oxide catalysts that are used in ozonation processes. In addition to these, granulated activated carbon (GAC), carbon black powder and graphite have been employed as catalysts or catalyst supports for Pt, Pd, Ru. Layered double hydroxides (LDH) are anionic clay minerals, which have also been used as ozonation catalysts. There are different types of clays, which are natural raw mixtures of clay minerals, volcanic ashes, silicas, carbonates, and miscellaneous and bentonite and one of the most studied clay mineral is montmorillonite (Shahidi et al. 2015).

References

Atalay S, Ersöz G (2015) Green chemistry for dyes removal from waste water: research trends and applications. In: Sharma SK (ed) Processes for removal of dyes from aqeous media. Scrivener Publishing LLC, Beverly, pp 83–117

Barb WG, Baxendale JH, George P, Hargrave KR (1949) Nature 163:692–694

Barb WG, Baxendale JH, George P, Hargrave KR (1951a) Trans Faraday Soc 47:462–500

Barb WG, Baxendale JH, George P, Hargrave KR (1951b) Trans Faraday Soc 47:591–616

Beltran FJ, Rivas FJ, Montero-de-Espinosa R (2005) Iron type catalysts for the ozonation of oxalic acid in water. Water Res 39:3553–3564

Bi DY (1999) Ind Catal 5:24–30

Fu D, Zhang F, Wang L, Yang F, Liang X (2015) Simultaneous removal of nitrobenzene and phenol by homogenous catalytic wet air oxidation. Chin J Catal 36(7):952–956

Gracia R, Aragües JL, Cortés S, Ovelleiro JL (1995) In: Proceedings of the 12th world congress of the international ozone association, vol 75. Lille

Jing G, Luan M, Chen T (2012) Progress of catalytic wet air oxidation technology. Arab J Chem (in press)

Karat I (2013) Advanced oxidation processes for removal of COD from pulp and paper mill effluents. A Technical, Economical and Environmental Evaluation. Industrial Ecology, Master of Science Thesis, Royal Institute of Technology

Kasprzyk-Hordern B, Ziólek M, Nawrocki J (2003) Catalytic ozonation and methods of enhancing molecular ozone reactions in water treatment. Appl Catal B Environ 46:639–669

Kim KH, Ihm SK (2011) Heterogeneous catalytic wet air oxidation of refractory organic pollutants in industrial wastewaters: a review. J Hazard Mater 186(1):16–34

Levec J (1997) Wet oxidation processes for treating industrial wastewaters. Chem Biochem Eng 11:47–58

Levec J, Pintar A (2007) Catalytic wet-air oxidation processes: a review. Catal Today 124:172–184

Lin SH, Ho SJ (1996) Appl Catal B Environ 9:133–147

Liotta LF, Gruttadauria M, DiCarlo G, Perrini G, Librando V (2009) Heterogeneous catalytic degradation of phenolic substrates: catalysts activity. J Hazard Mater 162:88–606

Liu J, Zhou L (1998) Water Purif Technol 65:6–10

Loures CCA, Alcântara MAK, Filho IHJ, Teixeira ACSC, Silva FT, Paiva TCB, Samanamud GRL (2013) Advanced oxidative degradation processes: fundamentals and applications. Int Rev Chem Eng 5(2):102–120

Luck F (1999) Wet air oxidation: past, present and future. Catal Today 53:81–91

Mandal A, Ojha K, Bhattacharjee S, De AK (2004) Removal of catechol from aqueous solution by advanced photo-oxidation process. Chem Eng J 102(2):203–208

Mota ALN, Albuquerque LF, Beltrame LTC, Chiavone-Filho O, Machulek A Jr, Nascimento CAO (2008) Advanced oxidation processes and their application in the petroleum industry: a review. Braz J Pet Gas 2(3):122–142

Munter R (2001) Advanced oxidation processes—current status and prospects. Proc Estonian Acad Sci Chem 50:59

Muruganandham M, Suri RPS, Jafari S, Sillanpää M, Lee GJ, Wu JJ, Swaminathan M (2014) Recent developments in homogeneous advanced oxidation processes for water and wastewater treatment 21

Navalon S, Alvaro M, Garcia H (2010) Heterogeneous Fenton catalysts based on clays, silicas and zeolites. Appl Catal B 99:1–26

Nawrocki J, Kasprzyk-Hordern B (2010) The efficiency and mechanisms of catalytic ozonation. Appl Catal B Environ 99:27–42

Neyens E, Baeyens J (2003) A review of classic Fenton's peroxidation as an advanced oxidation technique. J Hazard Mater 98:33–50

Pachhade K, Sandhya S, Swaminathan K (2009) Ozonation of reactive dye, Procion red MX-5B catalyzed by metal ions. J Hazard Mater 167:313–318

Pines DS, Reckhow DA (2002) Effect of dissolved cobalt(II) on the ozonation of oxalic acid. Environ Sci Technol 36:4046–4051

Pouran SR, Raman AAA, Dau WMAW (2014) Review on the application of modified iron oxides as heterogeneous catalysts in Fenton reactions. J Cleaner Prod 64:24–35

Rubin MB (2008) The history of ozone VI, Ozone of Silca Gel ("dry ozone"). Bull Hist Chem 33:68–75

Sable SS (2014) Development of novel catalytic materials for removal of emerging organic pollutants by advanced oxidation processes. Universitat Rovira I Virgili, Doctorate Thesis

Shahidi D, Roy R, Azzouz A (2015) Advances in catalytic oxidation of organic pollutants-prospects for thorough mineralization by natural clay catalysts. Appl Catal B Environ 174–175:277–292

Sharma S, Ruparelia JP, Patel ML (2011) A general review on advanced oxidation processes for waste water treatment. Institute of Technology, Nirma University, Ahmedabad, pp 382–481, 08–10 Dec 2011

Shiraga M, Kawabata T, Li D, Shishido T, Komaguchi K, Sano T, Takehira K (2006) Appl Clay Sci 33:247–259

Siitonen J (2007) Paperitehtaan Poistoveden Otsonointi. Bachelor Thesis, Lappeenrannan Technical University

Soon AN, Hameed BH (2011) Heterogeneous catalytic treatment of synthetic dyes in aqueous media using Fenton, and photo-assisted Fenton process. Desalination 269:1–16

Trapido M, Veressinina Y, Munter R, Kallas J (2005) Catalytic ozonation of m-dinitrobenzene. Ozone Sci Eng 27:359–363

Wang YZ, Li Z, Yin L, Hu KY (1993) Environ Chem 12:408–413

Wu CH, Kuo CY, Chang CL (2008) Homogeneous catalytic ozonation of C.I. Reactive Red 2 by metallic ions in a bubble column reactor. J Hazard Mater 154:748–755

Xiao H, Liu R, Zhao X, Qu J (2008) Effect of manganese ion on the mineralization of 2,4-dichlorophenol by ozone. Chemosphere 72:1006–1012

Yalfani MS (2011) New catalytic advanced oxidation processes for wastewater treatment. Doctoral Thesis, Universitat Rovira i Virgili

Zhao L, Ma J, Sunb Z, Zhai X (2008) Catalytic ozonation for the degradation of nitrobenzene in aqueous solution by ceramic honeycomb-supported manganese. Appl Catal B Environ 83:256–264

Zhao L, Ma J, Sunb Z, Zhai X (2009) Mechanism of heterogeneous catalytic ozonation of nitrobenzene in aqueous solution with modified ceramic honeycomb. Appl Catal B Environ 89:326–334

Chapter 4
Review on Catalysis in Advanced Oxidation Processes

Abstract Advanced oxidation processes (AOPs) constitute a promising technology for the treatment of wastewaters containing non-easily removable organic compounds. This chapter presents the case studies highlighting the application of advanced oxidation processes for wastewater treatment using the nanocatalysts, perovskite type catalysts or green catalysts. Some of the related studies in literature are summarized by providing the operating conditions and discussing the results gained.

4.1 Nanotechnology in Green Catalysis in Advanced Oxidation Process

Nanotechnology is fruitful area for green chemistry catalysis and due to the catalytic capabilities and efficiencies attained they are widely preferred to be used in AOPs.

Among the various AOPs, heterogeneous photocatalytic degradation of organic contaminants in the presence of nanostructured catalysts attained good efficiencies in degradation of organic compounds.

Especially textile dyes and dyeing agents represent one of the largest groups of organic molecules those are treated with photocatalytic degradation. Below some of the photocatalytic studies performed in literature are summarized:

Metal-oxide nanoparticles have been extensively utilized in photocatalytic oxidation processes, due to their interesting physicochemical properties (Das and Basu 2015; Chen et al. 2015; Sathishkumar et al. 2013). Among all semiconductor oxides, titanium dioxide is the most widely applied as photocatalyst since it is inexpensive, chemically stable, non-toxic and its photogenerated holes and electrons are highly oxidizing and reducing, respectively (Lucas et al. 2013; Litter 1999).

In 2015, Das and Basu investigated the photocatalytic treatment of textile effluent using titania–zirconia nano composite catalyst. They prepared the nano composite photocatalysts by sol–gel method. The TiO_2/ZrO_2 composite with a Zr/Ti mass ratio of 11.8 % was found to be an effective photocatalyst for treatment of textile wastewater and a large surface area (190 m^2/g), pore volume (0.72 ml/g) for 8 nm particle size. They reported that the degradation process is highly influenced by irradiation time, pH, effluent concentration, air flow rate and photocatalyst loading. The initial concentration of the effluent was also found to be an important parameter. It was seen that the photocatalyst can be reused up to five times. They assumed that this was basically a heterogeneous process and in the photocatalytic degradation of the dye the incident photons were absorbed by the photocatalyst. Considering the reaction is taking place on the catalyst surface the rate constant and the adsorption equilibrium constant decomposition of the dye were obtained from Langmuir–Hinshelwood model (Das and Basu 2015).

In a similar study which was performed in 2015, Chen et al. investigated the photodegradation of methyl blue (MB) in the presence of the nanosized Ag/CO_3O_4 composite photocatalysts with peroxymonosulfate (PMS) under visible light. Pure CO_3O_4 nanoflakes and Ag/CO_3O_4 nanocomposites with different Ag contents were prepared by hydrothermal method and co-precipitation method, respectively. They found that Ag doped on the CO_3O_4 surface decreased the band gap and suppressed the recombination of electron–hole pairs. Therefore, under visible light irradiation, the Ag/CO_3O_4 composites showed a higher photocatalytic efficiency for the degradation of MB than the pure CO_3O_4. In the range of the studied concentrations, pseudo zero-order kinetics was found to be the most suitable to describe the relationship between MB concentration and irradiation time in the initial photocatalytic dye disposal phase. The kinetic constant of pure CO_3O_4 nanoparticles for MB degradation was 0.0381 min^{-1}, which was far lower than that of Ag/CO_3O_4 composites. The enhanced photocatalytic activity of Ag/CO_3O_4 was mainly due to the coupling effect in the separation of photoexcited electron–hole pairs. The 3.06 % Ag-doped Ag/CO_3O_4 nanocomposite catalyst can achieve complete decolorization of 15 mg/L MB dye solution. They concluded that Ag/CO_3O_4 nanocomposite is a very promising catalyst for practical applications for wastewater treatment because of its high activity and stability (Chen et al. 2015).

The sono, photo and sonophotacatalytic degradation of Direct Blue 71 (DB71) accompanied by heterogeneous ZnO nanocatalysts were studied under the ultrasonic power. The effects of various conditions such as the dye concentration, pH and catalyst loading on the sonochemical degradation were investigated. Ertugay and Acar concluded that the ultrasonic degradation of dye was enhanced by the addition of the nanocatalyst. In addition, the effects of H_2O_2 on sonolysis and sonocatalytic oxidation were studied, and it was found that H_2O_2 enhanced the degradation of dye. Then, the effect of ZnO on photo and sonophotocatalytic oxidation was also investigated, and it was found that UV/ZnO combination was a more effective treatment on DB71 dyestuff (Ertugay and Acar 2014).

Photocatalytic degradation of methyl blue using Fe_2O_3/TiO_2 composite ceramics was realized by Li et al. in 2015. They used Fe_2O_3/TiO_2 nano powders as raw material for the preparation of Fe_2O_3/TiO_2 composite ceramics. They concluded that Fe_2O_3/TiO_2 ceramic showed a high photodegradation capability for methyl blue under either UV or visible light, the degradation rate could arrive to 83 and 88 % (Li et al. 2015).

Photocatalytic decolorization of malachite green dye by application of TiO_2 nanotubes were studied by Prado and Costa. The titania nanotubes were prepared in a hydrothermal system and they were characterized by scanning electronic microscopy (SEM), FT-IR, FT-Raman, and surface charge density by surface area analyzer. The prepared catalysts were used in malachite green dye degradation by phocatalytic oxidation. The photodegradation capacity of TiO_2 nanotubes was compared with TiO_2 anatase. Malachite dye was completely degraded in 75 and 105 min of reaction photocatalysed by TiO_2 nanotubes and TiO_2 anatase, respectively (Prado and Costa 2009).

However still in some cases, the recovery of the nanocatalysts from a heterogeneous suspension is one of the difficulties in wastewater treatment (Mourao et al. 2010). In order to overcome this bottleneck, recently, nanocatalysts with a combination of magnetic and non-magnetic particles were utilized in heterogeneous photocatalysis (Balu et al. 2011). By the application of a magnetic field, nanophotocatalysts having both magnetic and photocatalytic properties can be easily separated from the heterogeneous suspension. In recent years many studies have been focused on the synthesis of metal oxide nanoparticles because of their unusual optical, electrical and magnetic properties. Spinel ferrites, MFe_2O_4 (M: Mn, Co, Zn, Mg, etc.), are among the most important magnetic materials and one of the unique features of these materials is their superparamagnetism.

For example Sathishkumar et al. studied on the photocatalytic degradation of Reactive Red 120 (RR120) in aqueous solutions in the presence and absence of electron acceptors using magnetic $CoFe_2O_4/TiO_2$ nanocatalysts. Magnetic nanocatalysts were prepared by a co-precipitation method. The structural and elemental analyses confirmed the formation of $CoFe_2O_4$ and $CoFe_2O_4/TiO_2$ nanocatalysts and the specific atomic ratios of Ti, Co, Fe, and O in these catalysts. The presence of Co^{2+} and Fe^{3+} cations in its oxide forms on the surface of TiO_2 led to visible light absorption in the wavelength range 550–650 nm. The photocatalytic degradation of RR120 was studied by varying its concentration and the amount of nanocatalyst in order to attain a maximum degradation (Sathishkumar et al. 2013).

A TiO_2-coated magnetic nanomaterial, $Fe_3O_4@SiO_2@TiO_2$, was prepared, characterized and tested as photocatalyst in the degradation of Reactive Black 5 (RB5) dye by a UV-photocatalytic oxidation. In the study, two reference materials were also evaluated as photocatalysts in the same reaction: as-prepared non-magnetic TiO_2 and commercial TiO_2 from a Company (P25). FTIR and XRD data confirmed the presence of magnetite (Fe_3O_4) cores, silica and anatase in the magnetic nanomaterial. From the results they concluded that the TiO_2 shell in the $Fe_3O_4@SiO_2@TiO_2$ nanocomposite was photocatalytically active and promoted the dye oxidation in an extent almost similar to that achieved with the commercial P25. In addition to this as an another advantage of using, the magnetic nanocatalyst, the particles were efficiently separated from the reaction medium by magnetic decantation and could be reused three times with no decrease of the photocatalytic performance (Lucas et al. 2013).

Besides the dyeing agents, there are also various organic pollutants (especially phenol) that are treated under visible light with nanosized catalysts.

It is known that among the hazardous organics presented in wastewater, phenol and phenolic compounds are not only recalcitrant to natural degradation, but also toxic even at low concentration. For example, photochemical degradation of phenol solutions on Co_3O_4 nanorods with sulfate radicals was performed by Wang et al. In this study, Co_3O_4 nanorods were prepared by hydrothermal method followed and their catalytic performances in activation of peroxymonosulfate for phenol degradation were evaluated. It was found that Co_3O_4 nanorods showed a high catalytic activity in phenol oxidation with sulfate radicals. Co_3O_4 nanorods also possessed a high stability. The kinetic studies indicated that the heterogeneous catalytic system followed the first-order kinetics and the activation energy was 64.6 kJ/mol (Wang et al. 2015).

In another work, TiO_2 nanoparticles were prepared by sol–gel method and tested for the photocatalytic degradation of phenol with the addition of hydrogen peroxide. The kinetics of heterogeneous photocatalytic processes was studied by investigating the effect of operating parameters. The TiO_2 nanoparticles prepared in this study offered high photocatalytic activity to continuously produce the higher concentration of hydroxyl radicals (Ling et al. 2015).

Photocatalytic degradation of 2-chlorophenol (2-CP) was studied by Taleb using the photocatalyst chitosane/$CoFe_2O_4$ nanocomposite (CS/CF) under visible light. The nanocomposites were prepared by gamma irradiation cross-linking method. Their photocatalytic activity was tested for the degradation of 2-CP in aqueous medium using sunlight. The effect of different parameters such as catalyst concentration, 2-CP concentration and reaction pH on degradation was also examined. Taleb concluded that after the catalyst had been used 5 times repeatedly, the degradation rate was still above 80 %. It was also determined 2-CP that degradation obeys the pseudo-first-order kinetics for initial 2-CP concentrations between 25 and 100 mg/L, at 30 °C. The findings of the study clearly demonstrate that CS/CF nanocomposites could serve as convenient, effective, and recyclable photocatalysts for degradation of 2-CP from wastewater (Taleb 2014).

As mentioned in the previous sections, Fenton oxidation is one of the AOPs that can degrade recalcitrant compounds under mild conditions. The process can be performed both homogeneously and heterogeneously under various combinations. Recently, much attention has been paid to the Fenton-like processes, mainly focusing on the heterogeneous process in which nanosized catalysts are preferred.

Conventionally, metal oxides usually iron oxides are used as heterogeneous catalysts for Fenton oxidation system because of their abundance, easy separation and lower cost.

In 2015, Fathinia et al. prepared natural pyrite nanoparticles by high energy planetary ball milling as a nanocatalyst for heterogeneous Fenton process for decolorization of an organic dye, C.I. Acid Orange 7 (AO7), in aqueous solution. They analyzed the prepared catalysts by various characterization methods including XRD, SEM, EDX, FT-IR and BET analysis to assess the structural properties of pyrite nanoparticles. XRD results confirmed the crystallinity of the produced pyrite nanoparticles after high energy ball milling. SEM images exhibited nanoparticles with the size distribution of 20–100 nm. The maximum decolorization efficiency of 96.30 % was achieved at an initial AO7 concentration of 16 mg/L, catalyst dosage of 0.5 g/L, H_2O_2 concentration of 5 mmol/L and reaction time of 25 min. The results confirmed that pyrite nanoparticles can be used in five experimental cycles without considerable loss in catalytic activity. They concluded that pyrite nanoparticles could be considered as a novel heterogeneous Fenton process catalyst due to the significantly low price of natural pyrite in comparison to other synthetic catalysts (Fathinia et al. 2015).

In 2015 Hu et al. investigated removal of nitrobenzene from aqueous solution using nano-zero valent iron/granular activated carbon composite as Fenton-like catalyst. Granular activated carbon (GAC) supported nano-zero valent iron (nZVI) composite (nZVI/GAC) was prepared by adsorption–reduction method, and characterized by scanning electron microscopy, X-ray diffraction, X-ray photoelectron spectroscopy, and energy-dispersive X-ray spectroscopy. The catalytic degradation activity of the composite was evaluated to remove nitrobenzene pollutant by a heterogeneous Fenton-like system, and the initial pH value, nZVI/GAC dosage, and H_2O_2 concentration influencing on nitrobenzene removal were also investigated at room temperature. Experimental results showed that nZVI particle was uniformly dispersed over GAC matrix, and average particle size was 40–100 nm without agglomeration. The nZVI/GAC composite was very efficient in removing nitrobenzene with average percentage of more than 85 %. However, the removal rate of Fenton-like reaction was highly affected by pH value, H_2O_2 concentration, and nZVI/GAC dosage. The optimal reaction conditions were pH 4.0, 40 mg/L nitrobenzene, 5.0 mmol/L H_2O_2, and 0.4 g/L nZVI/GAC in the study (Hu et al. 2015).

Magnetite (Fe_3O_4), goethite ($\alpha FeOOH$), maghemite (γFe_2O_3) and hematite (αFe_2O_3) are widely used in nanosized heterogeneous catalysis processes and have been attractive alternatives for the Fenton treatment of wastewater. These materials can be easily removed from water by the simple use of a magnetic field.

In 2015, Nadejde et al. synthesized green Fenton-like magnetic nanocatalysts and investigated the characterization and catalytic application. They investigated the removal efficiency of micropollutants (Bisphenol A and Carbamazepine (CBZ), an anticonvulsant and mood-stabilizer) in the presence of magnetite-iron oxalate core–shell nanoparticle catalysts using heterogeneous Fenton-like oxidation and the effects of several parameters such as catalyst loading, H_2O_2 dosage, UV light and behavior of the mixed micropollutants on the oxidation of selected compounds. In the so called study, five iron oxalate core–shell magnetite nanoparticles catalysts were evaluated as magnetic heterogeneous Fenton catalysts. The samples were characterized by high resolution transmission electron microscopy and scanning electron microscopy. The EDX spectra and XRD patterns of coated catalysts were typical for pure magnetite. The observed FTIR peaks confirm the complexation of the iron oxalate by magnetite surface. The optimum experimental parameters were determined to be 1.0 g/L of catalysts, 10 mM H_2O_2, under UV irradiation. More than 99 % of both micropollutants were removed after 30 min of reaction time under the above experimental conditions (Nadejde et al. 2015).

The degradation of 2,4,6-trichlorophenol (TCP) was investigated by using magnetic nanoscaled Fe_3O_4/CeO_2 composite as a heterogeneous Fenton-like catalyst. The four process variables, pH, initial TCP concentration, Fe_3O_4/CeO_2 amount and H_2O_2 concentration, on TCP removal, mineralization and dechlorination were investigated by response surface methodology (RSM) using the central composite design (CCD). The optimum conditions were determined as pH: 2.0–2.1, TCP: 20–100 mg/L, Fe_3O_4/CeO_2: 1.5–2.5 g/L, and H_2O_2: 17–30 mM. The removal efficiency, mineralization and dechlorination rate of TCP were 99, 65 and 95 % after 90 min, respectively (Xu and Wang 2015).

Nanosized catalysts are also found some application in another advanced oxidation method which is wet air oxidation, too.

Low pressure catalytic wet air oxidation of aniline was investigated in a bubble reactor over nano Co_3O_4 (10 %wt)/CeO_2. The catalyst was prepared by sol–gel technology and characterized by using Scanning Electron Microscope, X-ray diffraction, nitrogen adsorption and thermogravimetric analysis techniques. The aim was to search for the conditions to destroy the aniline content by avoiding by production of byproducts such as ammonium, nitrate, and nitrite ions. The reaction was optimized at 0.5 g/L catalyst loading at 150 °C with a pressure of 4 atm, in 2 h with an air flowrate of 1.36 L/min. 35.15 % of aniline was removed and 14 % of the input nitrogen was converted into N_2 gas (Ersöz and Atalay 2010). The same authors also investigated the kinetic modeling of this process. Eight different models were proposed and tested. The tested models are presented in two sections which are Power-Law Based and Mechanism-Based Kinetic Expressions. The models were evaluated using the criterion on the minimization of the statistical parameter mean residual sum of squares (MRSS) and average experimental error. The models that led to thermodynamically inconsistent adsorption constants or negative kinetic parameters in the considered models were rejected. The oxidation reaction was best characterized by the Mars–van Krevelen model in which the reaction takes place through alternative oxidation and reduction of the catalyst surface, the surface oxidation being produced by molecular oxygen dissociatively adsorbed from the gas phase (Ersöz and Atalay 2011).

In another study, Ersöz and Atalay tested a different nanocatalyst, CuO/CeO_2 (10 % wt) and NiO/Al_2O_3 (10 % wt), on catalytic wet air oxidation of aniline. The experiments were again performed to investigate the effects of catalyst loading, temperature, reaction time, air flow rate, and pressure on aniline removal. The prepared catalysts seem to be active having an aniline removal of 45.74 % with CuO/CeO_2 and 41.90 % with NiO/Al_2O_3. The amount of N_2 formed was approximately the same for both of the catalysts (Ersöz and Atalay 2012).

Parvas et al. (2014) examined for catalytic wet air oxidation of phenol by non-noble metal Ni with different loadings was coated on precipitated nano CeO_2–ZrO_2 support. The structure of the nanocatalysts was determined by BET, FESEM, XRD, and FTIR analyses. They concluded that the catalytic wet air oxidation of phenol with different Ni loadings indicated improvement of phenol destruction at higher amounts of active phase. Removal of phenol increased with increasing catalyst loading from 4 to 9.0 g/l but further increase to 10 g/l declined the catalyst reactivity (Parvas et al. 2014).

Nanosized catalysts are also found application in other advanced oxidation processes, too.

In 2015, Taseidifar et al. investigated the production of nanocatalyst from natural magnetite by glow discharge plasma for enhanced catalytic ozonation of Basic Blue 3 (BB3) dye in aqueous solution. Cheap natural magnetite (NM) was modified with oxygen plasma owing to its cleaning effect by chemical etching and with argon plasma due to its sputtering effect resulting in more surface roughness. The performance of the plasma treated magnetites (PTMs) was higher than NM for treatment of Basic Blue 3 in catalytic ozonation (O_3/PTM). The properties of NM and the most efficient treated magnetite (PTM4) samples were characterized by X-ray diffraction, Fourier transform infrared spectroscopy, Brunauer–Emmett–Teller and scanning electron microscopy methods. The optimal values were found to be for operational parameters including ozone concentration (0.3 g/L), initial pH (6.7) and PTM4 dosage (600 mg/L). In the study they concluded that the PTM4 have the main advantages are environmentally friendly, they can be simply separated, negligible leached iron concentration, have successive reusability at milder pH and unaffected efficiency in the presence of inorganic salts (Taseidifar et al. 2015)

In 2015, Zhu et al. prepared Nickel cobalt oxide hollow nanosponges as advanced electrocatalysts for the oxygen evolution reaction. A class of novel nickel cobalt oxide hollow nanosponges (HNSs) was synthesized through a sodium borohydride reduction strategy. Significantly, taking advantage of the remarkable features of the novel materials, such as their porous and hollow nanostructures, and synergetic effects between their components, the optimized Ni–Co_2–O with spinel structure exhibit excellent catalytic activity, which makes them ideal candidates for the next generation oxygen evolution reaction catalysts. This class of nickel cobalt oxides can offer very attractive prospects and could be extended to electrochemical applications in various fields, such as supercapacitors, fuel cells, batteries and electrochemical sensors (Zhu et al. 2015).

In some cases, instead of using a single AOP, the combination of two or more AOPs in the presence of nanocatalysts has been found to be effective in enhancing the performance of individual processes as well as being cost-effective (Rey et al. 2011).

For instance, a novel coupled system using Co/TiO_2 (prepared using a sol–gel method) which combined two different heterogeneous advanced oxidation processes, sulfate radical based Fenton-like reaction and visible light photocatalysis, for degradation of organic contaminant, Rhodamine B (RhB) was performed. The Rhodamine B degradation rate and TOC removal were 100 and 68.1 % using the SR-Fenton/Vis-Photo combined process under ambient conditions, respectively. The authors reported that the hybrid system showed good catalytic stability and reusability, and almost no dissolution of Co^{2+} was found (Chen et al. 2014). In Table 4.1, some of the studies performend on AOPs using Nano Catalysts in literature are given.

Table 4.1 Literature survey on AOPs using nanocatalysts

Reference	Treatment method	Target compound	Catalyst	Reaction conditions/optimum conditions	Result
Das and Basu (2015)	Photocatalytic oxidation	Textile wastewater	TiO_2/ZrO_2 nano composite Zr/Ti mass ratio of 11.8 %	COD degradation increases up to catalyst loading was 500 mg/L and then it's decreases at 800 mg/L of catalyst concentration COD degradation is higher in acidic medium The optimum air flow rate is 0.3 L/min	TiO_2/ZrO_2 – an effective photocatalyst for treatment of textile wastewater The Langmuir–Hinshelwood based model
Chen et al. (2015)	Photocatalytic oxidation	Methyl blue (MB)	Nanosized Ag/Co_3O_4 composite	–	3.06 % Ag-doped Ag/Co_3O_4 can achieve complete decolorization of 15 mg/L MB dye solution
Ertugay and Acar (2014)	Photocatalytic oxidation Sono photocatalytic oxidation	Direct Blue 71 (DB71)	ZnO nanocatalysts	–	UV/ZnO combination was a more effective treatment on DB71 dyestuff
Li et al. (2015)	Photocatalytic oxidation	Methyl blue	Fe_2O_3/TiO_2 nano composite ceramics	MB (25 mg/L) in aqueous solution	Fe_2O_3/TiO_2 ceramic showed a high photodegradation capability under either UV or visible light, degradation rate arrive to 83, 88 %
Prado and Costa (2009)	Photocatalytic oxidation	Malachite dye	TiO_2 nanotubes	t = 75 and 105 min pH = 4	Completely degraded
Sathishkumar et al. (2013)	Photocatalytic oxidation	Reactive Red 120 (RR120)	Magnetic $CoFe_2O_4/TiO_2$ nanocatalysts	–	The synthesized nanocatalysts show higher efficiency towards the photocatalytic degradation of RR120 and the use of electron acceptors

(continued)

Table 4.1 (continued)

Reference	Treatment method	Target compound	Catalyst	Reaction conditions/optimum conditions	Result
Lucas et al. (2013)	Photocatalytic oxidation	Reactive Black 5 (RB5)	Magnetite (Fe_3O_4) cores, silica and anatase [$Fe_3O_4@SiO_2@TiO_2$] commercial P25	[RB5] = 50 mg/L in water; T = 20 °C; pH = 5.7;	$Fe_3O_4@SiO_2@TiO_2$ nanocomposite was photocatalytically active and promoted the dye oxidation in an extent almost similar to that achieved with the commercial P25
Wang et al., in press	Photocatalytic oxidation	Phenol and phenolic compounds	Co_3O_4 nanorods	[phenol]$_0$ = 20 mg/L, catalyst loading = 0.2 g/L and T = 25 °C	Co_3O_4 nanorods showed a high catalytic activity followed the first-order kinetics and the activation energy was 64.6 kJ/mol
Ling et al. (2015)	Photocatalytic oxidation	Phenol	TiO_2 nanoparticles	–	TiO_2 nanoparticles prepared in this study offered high photocatalytic activity
Taleb (2014)	Photocatalytic oxidation	2-chlorophenol (2-CP)	Chitosane/$CoFe_2O_4$ nanocomposite (CS/CF)	Initial 2-CP concentrations between 25 and 100 mg/l, 30 °C	Degradation rate above 80 % the pseudo-first-order kinetics
Fathinia et al. (2015)	Fenton oxidation	C.I. Acid Orange 7 (AO7)	Natural pyrite nanoparticles	An initial AO7 concentration of 16 mg/L, catalyst dosage of 0.5 g/L, H_2O_2 concentration of 5 mmol/L and reaction time of 25 min	Decolorization efficiency of 96.30 %
Hu et al. (2015)	Fenton oxidation	Nitrobenzene	Granular activated carbon (GAC) supported nano-zero valent iron (nZVI) composite (nZVI/GAC)	pH 4.0, 40 mg/L nitrobenzene, 5.0 mmol/L H_2O_2, and 0.4 g/L nZVI/GAC	Removal of nitrobenzene with average percentage of 85 %

(continued)

Table 4.1 (continued)

Reference	Treatment method	Target compound	Catalyst	Reaction conditions/optimum conditions	Result
Nadejde et al. (2015)	Fenton oxidation	Micropollutants	Magnetic nanocatalysts	1.0 g/L of catalysts, 10 mM H_2O_2, under UV irradiation. 30 min of reaction time	More than 99 % of both micropollutants were removed
Xu and Wang (2015)	Fenton oxidation	2,4,6-trichlorophenol (TCP)	Nanoscaled Fe_3O_4/CeO_2 composite	pH: 2.0–2.1, TCP: 20–100 mg/L, Fe_3O_4/CeO_2: 1.5–2.5 g/L, and H_2O_2: 17–30 mM	The removal efficiency, mineralization and dechlorination rate of TCP were 99, 65 and 95 % after 90 min, respectively
Ersöz and Atalay (2010), Ersöz and Atalay (2011)	CWAO	Aniline	Nano Co_3O_4 (10 % wt)/CeO_2	At 0.5 g/L catalyst loading at 150 °C with a pressure of 4 atm, in 2 h with an air flowrate of 1.36 L/min	35.15 % of aniline was removed and 14 % of the input nitrogen was converted into N_2 gas characterized by the Mars–van Krevelen model
Ersöz and Atalay (2012)	CWAO	Aniline	Nano CuO/CeO_2 (10 % wt) and nano NiO/Al_2O_3 (10 % wt),	CuO/CeO_2: 125 °C, 5 atm, 0.5 g/L NiO/Al_2O_3: 150 °C, 5 atm, 0.5 g/L	Aniline removal of 45.74 % with CuO/CeO_2 and 41.90 % with NiO/Al_2O_3. The amount of N_2 formed was approximately the same for both of the catalysts
Parvas et al. (2014)	CWAO	Phenol	Non-noble metal Ni with different loadings was coated on precipitated nano CeO_2–ZrO_2 support	–	Removal of phenol increased with increasing catalyst loading from 4 to 9.0 g/L
Taseidifar et al. (2015)	Catalytic ozonation	Basic Blue 3 (BB3)	Natural magnetite (NM) was modified with oxygen plasma	Ozone concentration (0.3 g/L), initial pH (6.7) and PTM4 dosage (600 mg/L)	The catalyst can be simple separated from the of the catalyst, negligible leached iron concentration, have successive reusability at milder pH and unaffected efficiency in the presence of inorganic salts

4.2 Perovskite Type Catalysts in Advanced Oxidation Process

Perovskite-type oxides are well-known functional materials with a wide range of applications, and are used as catalysts for advanced oxidation processes as environmental catalysts. Generally, an increase in the specific surface area of a perovskite-type oxide improves its catalytic activity. To date, several studies have been performed in literature. Among these some are given below:

Sannino et al., performed a study on photo-Fenton oxidation of acetic acid using structured catalysts. In the work photo-Fenton oxidation of acetic acid, was carried out on Fe-, Mn-, Co-, Ni-, Cu-based perovskites supported on cordierite monolith, in the presence or in the absence of low amounts of Pt. Homogeneous photo-Fenton reaction by ferrioxalate complex has been also performed for comparison. At first they prepared the structured catalysts: The honeycomb cordierite support was made by the extrusion of plastic paste followed by drying and calcination and then supported perovskites $LaMeO_3$ (Me = Mn, Co, Fe, Ni, Cu) were prepared by impregnation of thin wall of monolithic honeycomb cordierite support. The solutions of nitrate salts in the ethylene glycol with added citric acid were used. Finally they tested the efficiencies of the structured catalysts. The comparison of homogeneous and heterogeneous photo-Fenton oxidation indicated that the use of a heterogeneous structured catalyst greatly improves the total organic carbon (TOC) removal and leads to a more effective use of H_2O_2. They concluded that the structured catalysts allowed the operation in a wider pH range without the formation of sludge or significant metal leaching. Finally, they mentioned that $LaFeO_3$ and $Pt/LaMnO_3$ resulted as the best catalysts for the process in terms of reaction rate (Sannino et al. 2011).

Sannino et al., also performed another study in 2013, on mathematical modelling of the heterogeneous photo-Fenton oxidation of acetic acid on structured catalysts. This work focused on the development and the validation of a mathematical model for the UV-C photo-Fenton degradation of acetic acid using a $LaFeO_3$ heterogeneous structured catalyst with a monolithic structure. A combined study of the evolved gases and the liquid phase was conducted. The experimental results led to the identification of the main reactions that occur in the system: the complete mineralization of acetic acid by H_2O_2 due to the presence of the catalyst and the decomposition of H_2O_2 to water and O_2 in the homogeneous phase. A mathematical model was then developed using Eley–Rideal-type kinetics for the acetic acid consumption and first-order kinetics for the photolysis of hydrogen peroxide. The apparent kinetic constants were estimated, and the accuracy of the model was tested under different experimental conditions to evidence the predictive capability of the model. A very good agreement between the mathematical model calculations and the experimental data obtained in the presence of various H_2O_2 dosages was achieved. The results demonstrate that the continuous addition of H_2O_2 during the photo-Fenton reaction ensures the fruitful utilization of the produced hydroxyl radicals in the acetic acid oxidation reaction to minimize the oxidant consumption and maximize the TOC removal in a shorter irradiation time (Sannino et al. 2013).

Taran et al., reported the use of perovskite-like catalysts $LaBO_3$ (B = Cu, Fe, Mn, CO, Ni) for wet peroxide oxidation of phenol in 2016. The perovskite-like oxides were prepared by the Pecini method. The study showed the activity of only $LaCuO_3$ and $LaFeO_3$ perovskite-like catalysts, Cu-containing catalysts being more active, though Fe-containing being more stable. The leaching test proved the heterogeneous nature of the catalyst action. The long-term experiments revealed the acceptable stability of the $LaFeO_3$ catalyst. XRD studies of the used samples demonstrated the stability of the perovskite-like structure of the catalysts during the reaction (Taran et al. 2016).

In 2014 Rusevova et al. used $LaFeO_3$ (LFO) and $BiFeO_3$ (BFO) perovskites as nanocatalysts for degradation of contaminants with heterogeneous Fenton-like reactions. Nano-crystalline perovskites were synthesized by sol–gel method using citric acid as complexing agent. They studied the degradation of phenol and methyl tert-buthyl ether as contaminants. Degradation was examined at reaction temperatures 20, 40 and 60 °C at pH 7. In the experiments, a small increase in temperature lead to a strong acceleration of oxidation. Also, the slight decrease in pH value significantly enhanced the degradation of phenol with both, LFO and BFO. Increasing H_2O_2 and catalyst concentration lead to increasing concentration of ·OH, which lead to increasing degradation. However, at some point the catalyst surface or H_2O_2 can become a dominant sink of ·OH, than further increase of catalyst surface does not lead to increased contaminant degradation. Furthermore, they concluded that if the H_2O_2 concentration was too high, it could inhibit catalytic activity (Rusevova et al. 2014).

In Table 4.2, some of the studies performed on AOPs using Perovskite Type Catalysts in literature are given.

Table 4.2 Literature survey on AOPs using perovskite type catalysts

Reference	Treatment method	Target compound	Catalyst	Result
Sannino et al. (2011)	Photo-Fenton oxidation	Acetic acid	$LaMeO_3$ (Me = Mn, Co, Fe, Ni, Cu)	The structured catalysts allowed the operation in a wider pH range without the formation of sludge or significant metal leaching $LaFeO_3$ and $Pt/LaMnO_3$ resulted as the best catalysts

(continued)

Table 4.2   (continued)

Reference	Treatment method	Target compound	Catalyst	Result
Sannino et al. (2013)	Photo-Fenton oxidation	Acetic acid	LaFeO$_3$ heterogeneous structured catalyst	A mathematical model was then developed using Eley–Rideal-type kinetics for the acetic acid consumption and first-order kinetics for the photolysis of hydrogen peroxide
Taran et al. (2016)	Wet peroxide oxidation	Phenol	LaBO$_3$ (B = Cu, Fe, Mn, Co, Ni)	The activity of only LaCuO$_3$ and LaFeO$_3$ perovskite-like catalysts, Cu-containing catalysts being more active, though Fe-containing being more stable
Rusevova et al. (2014)	Fenton-like reactions	Phenol and methyl tert-buthyl ether	LaFeO$_3$ (LFO) and BiFcO$_3$ (BFO) perovskites as nanocatalysts	LaFeO$_3$ and BiFeO$_3$ showed high catalytic activity in phenol oxidation at pH 7 Excellent stability of BiFeO$_3$ in four re-use cycles

4.3   Greener Catalysts in Advanced Oxidation Process

The current focus on green and sustainable manufacturing in the chemical industry necessitates the replacement of conventional treatment processes by cleaner catalytic alternatives. Hence with the increasing importance being placed on environmental issues, more green chemical processes are urgently needed to meet the challenging green requirements, especially in the catalysis industry (Wen et al. 2014).

Recently, many studies have been focused on the design and synthesis of environmental friendly catalysts. Selection of environmental friendly catalyst should also take into account its reusability, ease of product recovery, low temperature and leaching effect of the metal.

It is well known that the countries which have large agricultural resources also produce huge amounts of wastes, that can be an environmental hazard or difficult to store. However, these agriwastes (rice husk, green tea leaves etc.) can also be considered as a source of low cost renewable raw materials that with the proper treatment can have applications in a wide range of processes.

Rice husks have no commercial value and are normally treated as waste thereby causing environmental pollution and disposal problems. Recently, efforts are being put into not only overcoming the pollution issues but also finding value addition to these wastes by using them as secondary resource materials. Recently, rice husk is used as precursors to produce activated carbon (AC). With porous structure, high surface area and low cost, AC has attracted considerable attention and has been widely used as catalyst support.

In literature there are many studies performed in advanced oxidation of various wastewaters using rice husk or rice husk ash as catalyst or activated carbon. For example, there are several photocatalytic studies that the efficiencies of the rice husks have been tested:

In 2015, Sinha and Ahmaruzzaman studied a novel green synthesis of gold nanorice and its utilization as a catalyst for the degradation of hazardous dye, Eosin Y, by photodegradation. A simple, green and template free method was developed for the production of rice shaped gold nanostructures using an aqueous extract of the egg shells of Anas platyrhynchos. The prepared nanoparticles were characterized by UV-visible, transmission electron microscopy (TEM), selected area electron diffraction pattern (SAED) and FT-IR studies. The TEM and SAED pattern confirmed the morphology, size and crystallographic structure of the synthesized gold nanorice. It was observed that the dye was degraded completely within 1 h and the percentage efficiency was found to be 96 % (Sinha and Ahmaruzzaman 2015).

The adsorption and photocatalytic studies of methylene blue were performed by Adam et al. using ceria and titania incorporated silica based catalyst prepared from rice husk. Rice husk silica based catalyst was synthesized via sol–gel method. The catalyst was used to study the adsorption and photodegradation of methylene blue (MB) under UV irradiation. The catalyst showed excellent adsorption capability with more than 99 % removal of MB from a 40 mg/L solution in just 15 min. It also decolorized an 80 mg/L MB solution under UV irradiation in 210 min, which was comparable with the commercialized pure anatase TiO_2 (Adam et al. 2013).

In 2013, Mohamed et al. prepared zeolite Y from rice husk ash encapsulated with $Ag-TiO_2$. They investigated the characterization and applications of this green catalyst in photocatalytic degradation of cyanide. A high level of efficiency in the photocatalytic reactions was achieved by increasing the surface area of the photocatalyst by supporting fine TiO_2 particles of zeolites. Ti-incorporated Y zeolite was prepared by an ion-exchange method, while Ag was immobilized on the encapsulated titanium via impregnation method. The produced samples were characterized using X-ray diffraction, ultraviolet and visible spectroscopy, photoluminescence emission spectra, scanning electron microscopy, and surface area measurement. Furthermore, the catalytic performances of Ti-Ag/NaY tests were carried out for degradation of cyanide using visible light. The results reveal a good distribution of Ag on the zeolite. Ag doping can eliminate the recombination of electron-hole pairs in the catalyst. These results demonstrate that the optimum weight% of Ag to Ti-NaY is 0.3 %; this weight% facilitates high performance by the photocatalyst, degrading 99 % of cyanide in a 100 mg/L solution in 60 min (Mohamed et al. 2013).

The use of rice husk-based catalyst for heterogeneous Fenton-like degradation of textile dyes is rarely reported. In particular, Daud and Hameed have reported on the use of a rice husk ash-based (RHA) catalyst for decolorization of Acid Red 1 by Fenton-like process (Daud and Hameed 2010) and Gan and Li have studied on Fenton-like degradation of RhB using an iron silica catalyst in which silica was extracted from rice husk (Gan and Li 2013). And in another study, Fenton-like oxidation of Reactive Black 5 dye was carried out using iron (III) impregnated on rice husk ash as heterogeneous catalyst. The catalyst was prepared by wet impregnation method and characterized. The effectiveness of this catalyst in degradation and decolorization of the dye, as well as the influence of reaction parameters on the catalytic activity was discussed. The effects of pH, the initial hydrogen peroxide concentration, the catalyst loading, and the temperature on the oxidative degradation and decolorization of Reactive Black 5 have been assessed. The best degradation efficiency (59.71 %) and decolorization efficiency (89.18 %) was obtained at temperature = 30 °C, pH = 3, $[H_2O_2]o$ = 4 mM, catalyst loading = 0.5 g/L for initial dye concentration of 100 mg/L (Ersöz 2014).

In 2015, Vinothkannan et al. studied one-pot green synthesis of reduced graphene oxide (RGO)/Fe_3O_4 nanocomposites and its catalytic activity toward methylene blue dye degradation. Environmental friendly Solanum trilobatum extract has been proposed for the simultaneous reduction of GO and Fe^{3+} ions for the preparation of RGO/Fe_3O_4 nanocomposite. The morphological and structural characterizations revealed that the strong interaction was exerted between Fe_3O_4 nanoparticles and RGO matrix. The well dispersed and uniform sized Fe_3O_4 nanoparticles anchored over RGO matrix increased the surface to volume ratio and active sites of RGO/Fe_3O_4 composite. The extended catalytic active sites completely degraded the MB dye within 12 min. The low cost and environmental benign one pot synthesis proposed for the RGO/Fe_3O_4 composite opened the fundamental studies on the active carbon supported transition metal oxide (Vinothkannan et al. 2015).

In 2015 Tahir et al. investigated the photo catalytic activity of green synthesized silver nanoparticles using Salvadora persica stem extract. A novel and ecofriendly procedure was developed for the synthesis of silver nanoparticles (AgNPs). The aqueous stem extract of Salvadora persica was used as reducing and capping agent. The formation of nanoparticles was observed at different temperatures and concentrations of the stem extract. The AgNPs were evaluated for photodegradation activity against methylene blue (MB) as an experimental substrate. Effect of various experimental conditions, such as catalyst amount, irradiation time and shape, size and dispersion of AgNPs were also investigated on the photo degradation of MB. The irradiation time experiment showed that photo degradation of MB was a rapid process and decomposed 96 % in 80 min. The strong activities of AgNPs confirmed the significant application in water purification by converting hazardous materials into non-hazardous one (Tahir et al. 2015).

In 2015, Zhou et al. investigated the preparation of magnetic carbon composites from peanut shells and its application as a heterogeneous Fenton catalyst in removal of methylene blue. Magnetic carbons were prepared from agricultural waste peanut shells and ferric ammonium oxalate via a simple impregnation and carbonization process. The obtained composites were characterized by element analysis, Mössbauer spectroscopy, X-ray photoelectron spectroscopy, scanning electron microscopy, X-ray diffraction, vibrating sample magnetometry and the Brunauer-Emmett-Teller surface area method, respectively. The magnetic carbon material was used as catalyst of heterogeneous Fenton reaction to remove methylene blue with the help of persulfate in waste water. The results indicated that both the removal rate and removal efficiency of this catalytic system are very excellent. The degradation efficiency was best (90 % within 30 min) using initial concentrations of 0.5 g/L persulfate and 40 mg/L methylene blue. The catalyst retained its activity after seven reuses, indicating its good stability and reusability (Zhou et al. 2015).

In another study, the removal of Procion Red MX-5B azo dye was investigated using several advanced oxidation methods in the presence of walnut shell based catalysts and comparatively biological oxidation in which the walnut shells were used as immobilizing agent for the microorganisms. The advanced oxidation methods comprising Fenton like oxidation, photocatalytic oxidation, photo-Fenton like oxidation and catalytic wet air oxidation (CWAO) were applied in the presence of iron on activated carbon catalysts (Fe/AC) containing 10 % iron by weight. In addition to Fe/AC, Fe-TiO$_2$ loaded on activated carbon catalysts (Fe-TiO$_2$/AC) were used in photocatalytic oxidation. (Fe-TiO$_2$)/Catalyst:1/10 by wt. and Fe/TiO$_2$:1/100 by wt. in the photocatalyst. In the biological oxidation, due to the oxidizing enzymes they secreted, the white rot fungi Phanerochaete chrysosporium and Trametes versicolor were selected to remove Procion Red MX-5B. When the advanced oxidation and biological oxidation methods tested in the study were compared, a much more rapid dye removal was accomplished at higher removal efficiencies by the application of photo-Fenton like and Fenton like oxidation methods. Advanced oxidation methods have also some advantages over the biological oxidation methods such as ease of handling the process and acceptable operating costs (Palas 2015; Palas et al. 2013a, b).

Application of advanced oxidation and novel biological oxidation methods for treatment of terephthalic acid plant wastewater in the presence of walnut shells based catalysts was studied. The target compounds to represent terephthalic acid wastewater were selected as terephthalic acid (TPA), benzoic acid (BA) and p-toluic acid (p-Tol). Degradation of BA by various advanced oxidation methods such as Fenton-like oxidation, catalytic wet air oxidation, photocatalytic oxidation, photo-Fenton-like oxidation and by biological oxidation was investigated. Degradations of terephthalic acid and p-toluic acid by photocatalytic and photo-Fenton-like oxidation were also investigated. The first part was preparation and characterization of metal doped walnut shells based catalysts. Iron (Fe) or iron-titanium dioxide (Fe-TiO$_2$ (1:99)) with a weight ratio of 10 % were doped to the activated carbon (AC) prepared from walnut shells by physical and different chemical activation methods such as acidic, neutral and basic activation. After metal doping Fe/AC and Fe-TiO$_2$/AC catalysts were obtained. In the second part of the study, first catalyst screening experiments were performed to each target compound for methods applied to determine the most efficient catalyst. Then parametric studies were performed to determine optimum values of operating conditions. Initial concentration was kept constant as 50 mg/L for each target compound. For benzoic acid, neutral Fe-TiO$_2$/AC catalyst was selected for photocatalytic and photo-Fenton-like oxidation and highest degradation efficiency for BA, % 95 degradation was accomplished by photo-Fenton-like oxidation in 80 min at optimum operating conditions. For terephthalic acid, neutral Fe-TiO$_2$/AC catalyst was determined from catalyst screening experiments for photocatalytic and photo-Fenton-like oxidation and highest degradation efficiency for TPA in the presence of this catalyst was accomplished with a degradation efficiency of 90 % in 120 min at optimum operating conditions. For p-toluic acid, neutral Fe-TiO$_2$/AC catalyst same as other target compounds was determined for photocatalytic and photo-Fenton-like oxidation. Highest degradation efficiency for p-Tol in the presence of this catalyst was accomplished with a degradation efficiency of 70 % in 60 min at optimum operating conditions. In the third part of the study, biological oxidation of benzoic acid in the presence of biocatalysts based on walnut shells was investigated. *P. chrysosporium* and *T. versicolor* were selected as microorganisms and immobilized on walnut shells. *P. chrysosporium* immobilized on walnut shells were determined as more successful in biocatalyst screening experiments and in presence of this biocatalyst, up to 90 % degradation was achieved at optimum conditions in 4 days. As a conclusion, photocatalytic oxidation and photo-Fenton-like oxidations were determined as the most promising methods with higher degradation efficiencies and degradation rates for all target compounds. Comparing biological oxidation, photocatalytic oxidation, and photo-Fenton-like oxidation of BA in which highest degradation efficiencies were obtained for BA, the same degradation efficiency (90–95 %) was achieved in two hours by photocatalytic oxidation and photo-Fenton-like oxidation where it was achieved in four days by biological oxidation (Tekin 2015; Tekin et al. 2013a, b).

In 2014 Muthirulan et al. developed green approach to fabrication of titania–graphene nanocomposites for the photodegradation of Acid Orange 7 dye under solar irradiation. High-resolution transmission electron microscope (HRTEM) and field emission scanning electron microscope (FESEM) analyses confirmed that the TiO$_2$ nanoparticles are uniformly decorated on the grapheme (GR) surface. Spectroscopic studies (FTIR, Raman, XRD, BET and UV-DRS) evinced that the optimal assembly and interfacial coupling between the GR sheets and TiO$_2$ in TiO$_2$–GR nanocomposite. Electrochemical studies showed that the TiO$_2$–GR nanocomposite possess greater redox activity and electrical conductivity. The photodegradation of AO7 under the solar irradiation followed the pseudo first-order kinetics according to the Langmuir–Hinshelwood model. Photodegradation studies revealed that TiO$_2$–GR nanocomposite possess greater degradation efficiency for AO7 dye compared to TiO$_2$, indicating that the electron transfer between TiO$_2$ and GR will greatly retard the recombination of photo-induced charge carriers and prolong electron lifetime, which contributes to the enhancement of photocatalytic performance (Muthirulan et al. 2014).

In 2013 Taherian et al. investigated the sono-catalytic degradation and fast mineralization of p-chlorophenol in the presence of $La_{0.7}Sr_{0.3}MnO_3$ as a nano-magnetic green catalyst (LSMO). Nanoparticles with a perovskite structure were prepared by a combination of ultrasound and co-precipitation method. The synthesized catalyst was characterized by X-ray diffraction, transmission electron microscopy, Fourier transform infrared spectroscopy. The catalytic performance of the catalyst was evaluated for the degradation of 4-chlorophenol in the presence and in the absence of ultrasound. The degradation has been studied at different temperatures, pH, catalyst dosage, and initial concentration of 4-chlorophenol. The results showed that the degradation efficiency was higher in the presence of ultrasound than its absence under the mild conditions. More than 88 % decrease in the concentration and 85 % decrease in the TOC for 4-chlorophenol could be achieved in a short time of sonication with respect to the conventional method. The catalyst without recalcination can be used successfully up to five consecutive cycles without any significant loss in activity in the presence and in the absence of ultrasound. In addition, the most important is the magnetic property of the nanoparticles which separated easily from aqueous solution by an external magnetic field (Taherian et al. 2013).

In 2014 Mehdi et al. prepared a green photocatalyst by immobilization of synthesized ZnO nanosheets on scallop shell for degradation of C.I. Acid Red 14 (AR14). The mean crystallite size of the ZnO nanosheets on scallop shell sample was about 15 nm. Degradation efficiency of AR14 by immobilized ZnO is more efficient than suspension form in identical conditions. Effect of operational parameters including nanocatalyst dosage, pH and initial dye concentration on the UV/ZnO-scallop process was studied. Kinetic of the photocatalytic process was explained in terms of the Langmuir–Hinshelwood model. The values of the kinetic rate and the Langmuir adsorption constants were determined as 0.104 1/(mg/L) and 0.413 (mg/L)/min, respectively. The prepared photocatalyst demonstrated the proper photocatalytic activity even after five successive cycles (Mehdi et al. 2014).

In another study, degradation of a food dye, tartrazine, was investigated by photocatalytic and photo-Fenton like oxidation processes. Hydroxyapatite catalysts were used which were prepared from egg shells. Fe^{3+} was loaded by wet impregnation to the hydroxyapatite at different weight percentages (1, 5 and 10 %). Firstly, the catalyst screening experiments were performed for both photocatalytic and photo-Fenton like oxidation to investigate the effect of the Fe^{3+} amount on the degradation and decolorization performance. Neither degradation, nor decolorization could be achieved in the photocatalytic oxidation for all catalysts, thus no parametric study was performed for this method. For photo-Fenton-like oxidation, the most efficient catalyst was determined as 1 % Fe-HA catalyst and further parametric study was carried out in the presence of this catalyst. In the parametric study, the optimum values of operating parameters such as operating catalyst loading, UV light intensity, initial H_2O_2 concentration, and pH were searched for. Additionally, the effect of the initial dye concentration was also investigated. The optimum values were determined as 0.1 g/L for catalyst loading, 6 W for UV light intensity, 4 mM for H_2O_2 concentration, and approximately 6 (natural pH of dye solution) for pH with initial dye concentration of 50 ppm. The maximum degradation and decolorization efficiencies obtained were 94 and 96 % at optimum operating conditions, respectively. By increasing the initial dye concentration, both degradation and decolorization efficiency decreased (Arlı et al. 2015).

In 2013, Shariffuddin et al. prepared green photocatalysts using hydroxyapatite derived from waste mussel shells for the photocatalytic degradation of a model azo dye wastewater. The study demonstrated for the first time the feasibility of utilizing waste mussel shells for the synthesis of hydroxyapatite, $Ca_{10}(PO_4)_6(OH)_2$ (HAP) to be used as a greener, renewable photocatalyst for recalcitrant wastewater remediation. HAP was synthesised from Perna canaliculus (green-lipped mussel) shells using a novel pyrolysis wet slurry precipitation process. The physicochemical properties of the HAP were characterized using X-ray diffraction (XRD), Fourier transform infrared (FTIR) spectroscopy and scanning electron microscopy (SEM). The HAP produced was of comparable quality to commercial (Sulzer Metco) HAP. The synthesised HAP had good photocatalytic activity, whereby methylene blue and its breakdown products were degraded with an initial rate of 2.5×10^{-8} mol/L/min. The overall azo dye degradation was nearly 54 % within 6 h and 62 % within 24 h in an oxygen saturated feed in a batch reactor using a HAP concentration of 2.0 g/L, methylene blue concentration of 5 mg/L, UV irradiation wavelength of 254 nm and a stirring speed of 300 rpm. They also performed a kinetic study. The kinetics were well described by three first order reactions in series, reflecting the reaction pathway from methylene blue to azo dye intermediates, then to smaller more highly oxidised intermediates and finally degradation of the recalcitrant (Shariffuddin et al. 2013).

In 2011 Shahwan et al. investigated the green synthesis of iron nanoparticles and their application as a Fenton-like catalyst for the degradation of aqueous cationic and anionic dyes. Iron nanoparticles were produced using extracts of green tea leaves (GT-Fe NPs). The materials were characterized using TEM, SEM/EDX, XPS, XRD, and FTIR techniques and were shown to contain mainly iron oxide and iron oxohydroxide. The obtained nanoparticles were then utilized as a Fenton-like catalyst for decolorization of aqueous solutions containing methylene blue (MB) and methyl orange (MO) dyes. The related experiments investigated the removal kinetics and the effect of concentration for both MB and MO. The results indicated fast removal of the dyes with the kinetic data of MB following a second order removal rate, while those of MO were closer to a first order removal rate. The loading experiments indicated almost complete removal of both dyes from water over a wide range of concentration, 10–200 mg/L. Compared with iron nanoparticles produced by borohydride reduction, GT-Fe nanoparticles demonstrated more effective capability as a Fenton-like catalyst, both in terms of kinetics and percentage removal (Shahwan et al. 2011).

In Table 4.3 some of the studies on green catalysts are given.

Table 4.3 Literature survey on AOPs using greener catalysts

Reference	Treatment method	Target compound	Catalyst	Reaction conditions/optimum conditions	Result
Ersöz (2014)	Fenton-like oxidation	Reactive Black 5 (RB5)	Iron over rice husk ash	pH 3 30 °C $[H_2O_2]_0 = 4$ mM $[Dye]_0 = 100$ mg/L Catalyst loading = 0.5 g/L	The best degradation efficiency % 59.71 Decolorization efficiency (89.18 %) Noticeably low the iron leaching
Tahir et al. (2015)	Photocatalytic degradation	Methylene blue	Silver nanoparticles (AgNPs) sythesized using *Salvadora persica* as reducing and capping agent	Under UV light irridiation Amount of 8 mg of AgNPs was added into 70 mL of MB solution (15 mg/L)	Degradation efficiency of 96 % in 80 min
Zhou et al. (2015)	Fenton oxidation	Methylene blue	Magnetic carbon composites from peanut shells	60 mg catalyst was added to 10 mL of Initial concentrations of 0.5 g L^{-1} persulfate and 40 mg L^{-1}	Degradation efficiency of 90 % within 30 min
Shariffuddin et al. (2013)	Photocatalytic degradation	Methylene Blue	Hydroxyapatite derived (HAP) from waste Mussel shells	Oxygen saturated feed in a batch reactor using a HAP concentration of 2.0 g/L, methylene blue concentration of 5 mg L^{-1}, UV irradiation wavelength of 254 nm and a stirring speed of 300 rpm	54 % within 6 h and 62 % within 24 h
Zeng et al. (2013)	Photocatalytic degradation	Acid Yellow (AY)	TiO₂-coated activated carbon surface (TAs) Activated carbon was produced from coco nut shell	The initial content of AY was varied from 6 to 15 mg/L The aerating rate of air is 38 mL/s Reacting temperature: 25 °C 40 mW/cm² UV light intensity	The rate constants k_{app} of AY degradation: 0.0049 min^{-1} for pure TiO₂, 0.0057 min^{-1} for P25, 0.0070 min^{-1} for the mixture of TiO₂ and active carbon, 0.0098 min^{-1} for TAs. The relation between the AY degradation rates follows the order of TiO₂ content from high to low: 4.2, 5.1, 3.0, 6.8, 1.6 %

(continued)

Table 4.3 (continued)

Reference	Treatment method	Target compound	Catalyst	Reaction conditions/optimum conditions	Result
Mahmoodi et al (2011)	Photocatalytic degradation	Basic Red 18 Basic Red 46	Titania over Canola hull based activated carbon support (2 % wt. of AC)	7.68 mM of H_2O_2 0.128 mM of initial Basic Red 18 concentration, 0.120 mM of initial Basic Red 46 concentration, pH = 5.6	In 150 min, 81.2 and 76.2 % of COD removal for Basic Red 18 and Basic Red 46, respectively
Daud and Hameed (2010)	Fenton-like oxidation	Acid Red 1 (AR1)	Iron-rice husk ash	0.070 wt % of iron (III) oxide loading on RHA, dosage of catalyst = 5.0 g/L, initial pH = 2.0, $[H_2O_2]o$ = 8 mM, $[AR1]o$ = 50 mg/L at temperature 300 °C	96 % decolorization efficiency of AR1 was achieved within 120 min
Ramirez et al. (2007)	Fenton-like oxidation	Orange II (OII)	Activated carbon/Fe, Carbon aerogel/Fe, (Olive seed based activated carbon)	pH 3, 0.2 g/L catalyst amount, 6 mM H_2O_2 concentration 30 °C	Removal of TOC, and iron leaching, 94.6, 23 and 7.8 % respectively. The same values for carbon aerogel/Fe catalyst were 55, 58.8 and 10.0 % respectively

References

Adam F, Muniandy L, Thankappan R (2013) Ceria and titania incorporated silica based catalyst prepared from rice husk: adsorption and photocatalytic studies of methylene blue. J Colloid Interface Sci 406:209–216

Arlı E, Elçioğlu Yİ, Tekin G, Ersöz G, Atalay S (2015) Hydroxyapatite catalyst derived from egg shells for degradation of food dyes, diploma project, department of chemical engineering. Ege University, Bernova

Balu AM, Baruwati B, Serrano E, Cot J, Martinez JG, Varma RS, Luque R (2011) Magnetically separable nanocomposites with photocatalytic activity under visible light for the selective transformation of biomass-derived platform, molecules. Green Chem 13:2750–2758

Chen Q, Ji F, Guo Q, Fan J, Xu X (2014) Combination of heterogeneous Fenton-like reaction and photocatalysis using Co–TiO$_2$ nanocatalyst for activation of KHSO$_5$ with visible light irradiation at ambient conditions. J Environ Sci 26(12):2440–2450

Chen G, Si X, Yu J, Bai H, Zhang X (2015) Doping nano-Co$_3$O$_4$ surface with bigger nanosized Ag and its photocatalytic properties for visible light photodegradation of organic dyes. Appl Surf Sci 330:191–199

Das L, Basu JK (2015) Photocatalytic treatment of textile effluent using titania–zirconia nano composite catalyst. J Ind Eng Chem 24:245–250

Daud NK, Hameed BH (2010) Decolorization of acid Red 1 by Fenton-like process using rice husk ash-based catalyst. J Hazard Mater 176:938–944

Ersöz G (2014) Fenton-like oxidation of Reactive Black 5 using rice husk ash based catalyst. Appl Catal B 147:353–358

Ersöz G, Atalay S (2010) Low-pressure catalytic wet air oxidation of aniline over Co$_3$O$_4$/CeO$_2$. Ind Eng Chem Res 49:1625–1630

Ersöz G, Atalay S (2011) Kinetic modeling of the removal of aniline by low-pressure catalytic wet air oxidation over a nanostructured Co$_3$O$_4$/CeO$_2$ catalyst. Ind Eng Chem Res 50:310–315

Ersöz G, Atalay S (2012) Treatment of aniline by catalytic wet air oxidation: comparative study over CuO/CeO$_2$ and NiO/Al$_2$O$_3$. J Environ Manage 113:244–250

Ertugay N, Acar FN (2014) The degradation of direct blue 71 by sono, photo and sonophotocatalytic oxidation in the presence of ZnO nanocatalyst. Appl Surf Sci 318:121–126

Fathinia S, Fathinia M, Rahmani AA, Khataee A (2015) Preparation of natural pyrite nanoparticles by high energy planetary ball milling as a nanocatalyst for heterogeneous Fenton process. Appl Surf Sci 327:190–200

Gan PP, Li SFY (2013) Efficient removal of Rhodamine B using a rice hull-based silica supported iron catalyst by Fenton-like process. Chem Eng J 229:351–363

Hu S, Yao H, Wang K, Lu C, Wu Y (2015) Intensify removal of nitrobenzene from aqueous solution using nano-zero valent iron/granular activated carbon composite as fenton-like catalyst. Water Air Soil Pollut 226:155

Li R, Jia Y, Bu N, Wu J, Zhen Q (2015) Photocatalytic degradation of methyl blue using Fe$_2$O$_3$/TiO$_2$ composite ceramics. J Alloy Compd 643:88–93

Ling H, Kim K, Liu Z, Shi J, Zhu X, Huang J (2015) Photocatalytic degradation of phenol in water on as-prepared and surface modified TiO$_2$ nanoparticles. Catalysis Today (In press)

Litter M (1999) Appl Catal B 23:89–114

Lucas MS, Tavares PB, Peres JA, Faria JL, Rocha M, Pereira C, Freire C (2013) Photocatalytic degradation of Reactive Black 5 with TiO2-coated magnetic nanoparticles. Catal Today 209:116–121

Mahmoodi NM, Arami M, Zhang J (2011) Preparation and photocatalytic activity of immobilized composite photocatalyst (titania nanoparticle/activated carbon). J Alloy Compd 509:4754–4764

Mehdi SS, Alireza K, Behrouz V, Sang WJ, Sevda F (2014) Preparation of a green photocatalyst by immobilization of synthesized ZnO nanosheets on scallop shell for degradation of an Azo Dye. Curr Nanosci 10(5):684–694(11)

Mohamed RM, Mkhalid IA, Salam MA, Barakat MA (2013) Zeolite Y from rice husk ash encapsulated with Ag-TiO$_2$: characterization and applications for photocatalytic degradation catalysts. Desalination Water Treat 51:7562–7569

Mourao HAJL, Malagutti AR, Ribeiro C (2010) Synthesis of TiO$_2$-coated CoFe$_2$O$_4$, photocatalysts applied to the photodegradation of atrazine and rhodamine B. Appl Catal A Gen 382:284–292

Muthirulan P, Devi CN, Sundaram MM (2014) A green approach to the fabrication of titania–graphene nanocomposites: insights relevant to efficient photodegradation of Acid Orange 7 dye under solar irradiation. Mater Sci Semicond Process 25:219–230

Nadejde C, Neamtu M, Hodoroaba V-D, Schneider RJ, Paul A, Ababei G, Panne U (2015) Synthesis, characterization and catalytic application. Appl Catal B 176–177:667–677

Palas B (2015) Azo dye degradation by advanced oxidation methods using metal impregnated organic waste catalyst and comparatively biological oxidation. Master of Science Thesis, Ege University

Palas B, Ersöz G, Atalay S (2013a) Catalytic wet air oxidation of Procion Red MX-5B using iron impregnated chemically activated walnut shells as a catalyst. Paper presented at the second international conference on water, energy and environment (ICWEE'2013), Kuşadası, Türkiye, 21–24 Eylül, 2013

Palas B, Ersöz G, Atalay S (2013b) Fenton like oxidation of procion Red MX-5B using metal impregnated organic waste catalyst, Journal of Selçuk University Natural and Applied Science, ICOEST Conf. 2013 (Part 2):160–170

Parvas M, Haghighi M, Allahyari S (2014) Catalytic wet air oxidation of phenol over ultrasound-assisted synthesized Ni/CeO$_2$–ZrO$_2$ nanocatalyst used in wastewater treatment. Arab J Chem

Prado AGS, Costa LL (2009) Photocatalytic decouloration of malachite green dye by application of TiO$_2$ nanotubes. J Hazard Mater 169(1–3):297–301

Ramirez JH, Maldonado-Hodar FJ, Perez-Cadenas AF, Moreno-Castilla C, Costa CA, Madeira LM (2007) Azo-dye Orange II degradation by heterogeneous Fenton-like reaction using carbon-Fe catalysts. Appl Catal B 75:312–323

Rey A, Carbajo J, Adán C, Faraldos M, Bahamonde A, Casas JA et al (2011) Improved mineralization by combined advanced oxidation processes. Chem Eng J 174(1):134–142

Rusevova K, Köferstein R, Rosell M, Richnow HH, Kopinke F-D, Georgi A (2014) LaFeO$_3$ and BiFeO$_3$ perovskites as nanocatalysts for contaminant degradation in heterogeneous fenton-like reactions. Chem Eng J 239:322–331

Sannino D, Vaiano V, Ciambelli P, Isupova LA (2011) Structured catalysts for photo-Fenton oxidation of acetic acid. Catal Today 161:255–259

Sannino D, Vaiano V, Ciambelli P, Isupova LA (2013) Mathematical modelling of the heterogeneous photo-Fenton oxidation of acetic acid on structured catalysts. Chem Eng J 224:53–58

Sathishkumar P, Mangalaraja RV, Anandan S, Ashokkumar M (2013) CoFe$_2$O$_4$/TiO$_2$ nanocatalysts for the photocatalytic degradation of Reactive Red 120 in aqueous solutions in the presence and absence of electron acceptors. Chem Eng J 220:302–310

Shahwan T, Sirriah SA, Nairat M, Boyacı E, Eroglu AE, Scott TB, Hallam KR (2011) Green synthesis of iron nanoparticles and their application as a Fenton-like catalyst for the degradation of aqueous cationic and anionic dyes. Chem Eng J 172:258–266

Shariffuddin JH, Jones MI, Patterson DA (2013) Greener photocatalysts: hydroxyapatite derived from waste mussel shells for the photocatalytic degradation of a model azo dye wastewater. Chem Eng Res Des 91:1693–1704

Sinha T, Ahmaruzzaman M (2015) A novel green and template free approach for the synthesis of gold nanorice and its utilization as a catalyst for the degradation of hazardous dye. Spectrochim Acta Part A Mol Biomol Spectrosc 142:266–270

Taherian S, Entezari MH, Ghows N (2013) Sono-catalytic degradation and fast mineralization of p-chlorophenol: La$_{0.7}$Sr$_{0.3}$MnO$_3$ as a nano-magnetic green catalyst. Ultrason Sonochem 20:1419–1427

Tahir K, Nazir S, Li B, Khan AU, Khan ZUH, Ahmad A, Khan FU (2015) An efficient photo
    catalytic activity of green synthesized silver nanoparticles using Salvadora persica stem extract.
    Sep Purif Technol 150:316–324
Taleb MFA (2014) Adsorption and photocatalytic degradation of 2-CP in wastewater onto
    CS/CoFe$_2$O$_4$ nanocomposite synthesized using gamma radiation. Carbohydr Polym 114:65–72
Taran OP, Ayusheeva AB, Ogorodnikova AL, Prosvirin IP, Isupova LA, Parmon VN (2016)
    Perovskite-like catalysts LaBO3(B = Cu, Fe, Mn Co, Ni) for wet peroxide oxidation of phenol.
    Appl Catal B 180:86–93
Taseidifar M, Khataee A, Vahid B, Khorram S, Jood SW (2015) Production of nanocatalyst from
    natural magnetite by glow discharge plasma for enhanced catalytic ozonation of an oxazine dye
    in aqueous solution. J Mol Catal A Chem 404–405:218–226
Tekin G (2015) Application of advanced oxidation and novel biological methods for the treatment
    of terephthalic acid plant wastewater. Master of Science Thesis, Ege University
Tekin G, Ersöz G, Atalay S (2013a) Fenton-like oxidation of benzoic acid using Fe^{3+} doped basic
    activated walnut shells as a catalyst. Paper presented at the second international conference on
    water, energy and environment (ICWEE'2013), Kuşadası, Türkiye, 21–24 Eylül, 2013
Tekin G, Ersöz G, Atalay S (2013b) Fenton-like oxidation of benzoic acid using Fe^{3+} doped
    physically activated walnut shells as a catalyst, Journal of Selçuk University Natural and
    Applied Science, ICOEST Conf. 2013 (Part 2):130–137
Vinothkannan M, Karthikeyan C, Gnana kumar G, Kim AR, Yoo DJ (2015) One-pot green
    synthesis of reduced graphene oxide (RGO)/Fe$_3$O$_4$ nanocomposites and its catalytic activity
    toward methylene blue dye degradation. Spectrochim Acta Part A Mol Biomol Spectrosc
    136:256–264
Wang Y, Zhou L, Duan X, Sun H, TE Lee, Jin W, Wang S (2015) Photochemical degradation of
    phenol solutions on Co$_3$O$_4$ nanorods with sulfate radicals. Catal Today
Wen C, Yin A, Dai W-L (2014) Recent advances in silver-based heterogeneous catalysts for green
    chemistry processes. Appl Catal B 160–161:730–741
Xu L, Wang J (2015) Degradation of 2,4,6-trichlorophenol using magnetic nanoscaled Fe$_3$O$_4$/
    CeO$_2$ composite as a heterogeneous Fenton-like catalyst. Sep Purif Technol 149:255–264
Zeng M, Li Y, Ma M, Chen W, Li L (2013) Photocatalytic activity and kinetics for acid yellow
    degradation over surface composites of TiO$_2$-coated activated carbon under different
    photocatalytic conditions. Trans Nonferrous Met Soc China 23(4):1019–1027
Zhou L, Ma J, Zhang H, Shao Y, Li Y (2015) Fabrication of magnetic carbon composites from
    peanut shells and its application as a heterogeneous Fenton catalyst in removal of methylene
    blue. Appl Surf Sci 324:490–498
Zhu C, Wen D, Leubner S, Oschatz M, Liu W, Holzschuh M, Simon F, Kaskel S, Eychmuller A
    (2015) Nickel cobalt oxide hollow nanosponges as advanced electrocatalysts for the oxygen
    evolution reaction. Chem Commun 51:7851–7854

Chapter 5
Concluding Remarks

Abstract This chapter concludes this book by presenting the main information on recent advances in the mentioned advanced oxidation processes (AOPs) for wastewater treatment and on the novel catalysts used in the so called processes. The authors discuss the Fundamentals of AOPs, the application of the AOPs to the wastewater treatment and the parameters affecting the process performance. They present information on green chemistry and catalysts in Advanced Oxidation Processes.

This chapter concludes this book by presenting the main findings. The main findings are:

- Organic contaminants in water remains being a great concern to the environment especially those refractory and/or non-biodegradable pollutants that can hardly be treated by conventional methods.
- Advanced oxidation processes provide an efficient environmental friendly process alternative for treating wastewater contaminated with toxic organic compounds.
- Different AOP techniques have been developed thus allowing to make choices the most appropriate for the specific treatment problems. Major attention should be devoted by the engineers to fill some specific gap which exists for these techniques in the areas such as identification of reaction intermediates, development of rate expressions based on established reaction mechanisms, catalyst selection and criteria for cost effectiveness and maximum destruction efficiency.
- Among the AOPs, the most commonly used methods seem to be Fenton and Photofenton Oxidations.
- Mostly AOPs are used in, the treatment of dye containing wastewaters.
- Green chemistry is of great concern—manufacturing in a manner that is sustainable, safe, and non-polluting and that consumes minimum amounts of materials and energy while producing little or no waste material.

© The Author(s) 2016
S. Atalay and G. Ersöz, *Novel Catalysts in Advanced Oxidation of Organic Pollutants*, SpringerBriefs in Green Chemistry for Sustainability, DOI 10.1007/978-3-319-28950-2_5

- Increasing attention is being directed towards the development of novel catalytic systems with high performances from the viewpoints of environment-friendly green process, economical efficiency and minimum consumption of resources.
- Among the catalyst categories, it is found that novel catalysts: nanocatalysts, perovskite type catalysts and greener catalysts are the most preferred in advanced oxidation processes.
- When compared, the confirmation and optimization of the efficiency of the perovskite oxides catalysts and the degradation pathway for the wastewater treatment has received less attention in the literature.
- Mainly, the agriwastes (rice husk, green tea leaves etc.) are considered as a source of low cost renewable raw material for green catalyst preparation.

Printed in the United States
By Bookmasters